A Treasure Hunting Text

Ram Publications
Hal Dawson, Editor

Ghost Town Treasures: Ruins, Relics and Riches
Clear explanations on searching ghost towns and deserted structures; which detectors to use and how to use them.

Real Gold in Those Golden Years
Prescription for happier, more satisfying life for older men and women through metal detecting hobby, ideally suited to their lifestyles.

Let's Talk Treasure Hunting
Ultimate "how-to" book of treasure hunting — with or without a metal detector; describes all kinds of treasures and tells how to find them.

The New Successful Coin Hunting
The world's most authoritative guide to finding valuable coins, totally rewritten to include instructions for 21st Century detectors.

Modern Metal Detectors
Comprehensive guide to metal detectors; designed to increase understanding and expertise about all aspects of these electronic marvels.

Treasure Recovery from Sand and Sea
Step-by-step instructions for reaching the "blanket of wealth" beneath sands nearby and under the world's waters; rewritten for the 90's.

You Can Find Gold...With a Metal Detector
Explains in layman's terms how to use a modern detector to find gold nuggets and veins; includes instructions for panning and dredging.

Gold of the Americas
A history of gold and how the precious metal helped shape the history of the Americas; filled with colorful vignettes and stories of bravery and greed.

Gold Panning is Easy
Excellent field guide shows the beginner exactly how to find and pan gold; follow these instructions and perform as well as any professional.

Buried Treasures You Can Find
Complete field guide for finding treasure; includes state-by-state listing of thousands of sites where treasure is believed to exist.

New World Shipwrecks: 1492-1825
Comprehensive guidebook lists more than 4,000 shipwrecks; tells how to locate a sunken vessel and how to explore it.

Sunken Treasure: How to Find It
One of the world's foremost underwater salvors shares a lifetime's experience in locating and recovering treasure from deep beneath the sea.

Treasure Hunting

for Fun and Profit

Charles Garrett

ISBN 0-915920-90-5
Library of Congress Catalog Card No. 96-68435
Treasure Hunting for Fun and Profit
© Copyright 1997
Charles Garrett

Second Printing, November 1998

For FREE listing of related treasure hunting books write:

Ram Publishing Company

P.O. Box 38649   •   Dallas, TX 75238

Contents

By Charles Garrett

Treasure Hunting Texts
    Ghost Town Treasures: Ruins, Relics and Riches
    Let's Talk Treasure Hunting
    Real Gold in Those Golden Years
    The New Successful Coin Hunting
    The New Modern Metal Detectors
    Treasure Recovery from Sand and Sea
    Treasure Hunting Pays Off
    Treasure Hunting Secrets

With Roy Lagal
    Find Gold with a Metal Detector
    Modern Treasure Hunting
    Modern Electronic Prospecting

True Treasure Tales
    The Secret of John Murrell's Vault
    The Missing Nez Perce Gold

The Author

*T*he name of Charles Garrett ranks high on any list of individuals who have pioneered the development and use of metal detectors...whether for discovery of treasure...for security...or, for any other reason.

Charles Garrett did not set out to become the world's leading manufacturer of metal detection equipment. He prepared himself well, however, to become one of the world's foremost treasure hunters. After graduation from Lufkin High School and service in the U.S. Navy during the Korean conflict, he earned an electrical engineering degree from Lamar University in Beaumont and began his business career in Dallas with Texas Instruments and Teledyne Geotech.

Nearly four decades ago, then, Mr. Garrett was a young electrical engineer deeply engrossed in development of systems and equipment required by America's fledgling space effort. In devoting himself to his lifetime hobby of treasure hunting, however, he also designed and built metal detectors in his spare time. This avocation soon became a career when he and his wife Eleanor founded Garrett Electronics to produce his inventions.

Today, the name Garrett stands as a synonym for the treasure hunting metal detector. Mr. Garrett has become recognized as an unofficial spokesman for the hobby of treasure hunting and the metal detecting industry through a long list of honors, personal appearances and books. He is the author of several major works which have been accepted as veritable "texts" for treasure hunting. His expertise has also carried him into the allied fields of security screening and crime scene investigation. Garrett is now the world's foremost manufacturer of security metal detection equipment. Its famed Magnascanner and Super Scanner instruments protect air travelers all over the world and have been honored as the choice to safeguard historical and cultural treasures, Olympic athletes, presidents and kings.

He is married to the former Eleanor Smith of Pennington, TX, who has played a key role in the growth of Garrett Electronics. They have two sons and a daughter.

As a graduate engineer and a businessman, Mr. Garrett introduced discipline to the manufacture of metal detectors. He has generally raised the standards of metal detecting everywhere, while the hobby has grown from a haphazard pastime to almost a science.

Garrett quality is known throughout the world. From the beginning, Charles Garrett vowed "to practice what I preach" — in other words, to test his equipment in the field...to insure it will *work* for customers regardless of ground conditions and environment. Thus, with a metal detector of his own design he has searched for and found treasure on every continent except Antarctica. He has also scanned under lakes, seas and oceans of the world.

Hal Dawson
Editor, Ram Publishing

Dallas, Texas
Spring 1997

2

Coins Just Waiting
Under the Saloon Floor

*I*t was dark, but Rickey Kirk knew the old building had once been a saloon decades ago. With renovation in progress, he asked permission to scan the ground after flooring was torn out.

"Sure, go ahead...we won't be ready to start putting in the new floor for two or three hours."

Two or three hours! Although the area to be scanned amounted to little more than 2,000 square feet, Rickey's impulse was to race around hurriedly waving his searchcoil in the dark. But, the young Texan remembered the treasure books he'd read and curbed this urge. Instead, he simply switched on his Garrett and began a methodical search.

He had trained himself well in the hobby of metal detecting, and he relied on this detector. Also, he trusted his judgment that treasure was there, and he knew that he could find it. This was what he had been waiting for...what treasure hunting is all about!

Before workers returned, Rickey had found 20 old coins and a saloon token. Among these was an 1845 dime and a 1941 Mercury dime. Many of the coins were half-dimes, once the price of a beer. Rickey Kirk of Bonham, Texas, was successful because he made certain he was in the right place at the right time with a quality detector. Valuable old coins were his reward!

Introduction

*T*hat's a good question! Let me tell you about treasure hunting. Let me tell you about the fun, the excitement, the genuine pleasure that can come from finding coins, jewelry, relics, gold nuggets and countless other valuable items. And, I'm *not* going to tell you about "metal detecting." In fact, with today's modern instruments you need very little knowledge of "metal detecting" in order to find treasure. Touch just one control on a modern computerized detector and begin hunting. It's that simple. Don't believe anyone who tells you differently!

It wasn't always so. Not too many years ago using a metal detector effectively required specialized knowledge. You gained this expertise either through training or through hard work in the field, or perhaps a combination of the two. Some of my early books sought to provide such knowledge. Thus, they were essentially extensions of metal detector instruction manuals. These books, thankfully, are out of print.

A simpler approach is now welcomed by all. This *Treasure Hunting Text* provides all the information you'll ever need on the basic techniques of using a metal detector. It also includes suggestions about research that can help you select and discover various types of treasure, as well as methods for recovering them. More importantly, however, the book seeks to describe the real enjoyment that can come from finding treasure and to explain how you can easily achieve such pleasure. Moreover, this book is written especially for men and women, boys and girls, who have never hunted with a metal detector.

I have written many books, articles, training manuals and the like that deal with all aspects of treasure hunting with a metal detector. Most of these pertain to a single facet of the hobby, such as coin hunting, searching for gold and ghost towning, to name but three of my most popular books. This one, however, is written to present an overview of treasure hunting and to provide a basic knowledge of all aspects of the hobby.

Treasure Hunting for Fun and Profit is divided into three sections. Chapters 1 and 2 explain the hobby and give you *all* the information you need to begin hunting. That's right — after reading these two chapters you'll be ready to hunt with a computerized detector. Chapters 3 through 12 take you deeper into the hobby with discussions of hunting for various types of targets, which will enable you to specialize in as many areas as you desire. These chapters also include information on where to hunt, clothing and equipment, health and safety, special advantages of the hobby for children and older individuals and a brief discussion of laws affecting treasure hunting.

Finally, Chapters 13 and 14 will provide you information that will enable you to hold your own in any discussion about hunting for treasure with a metal detector. I'm tempted to say that these final two chapters offer a true "course" in metal detecting. But, if you're like me you've passed all the courses you intend to take and don't plan on any others. So, let's just say that these final chapters close out a book that contains all the knowledge you'll ever need to become an expert treasure hunter.

I should know because I've been a treasure hunter all my life. For half a century now I've used a metal detector, andI can assure you that finding treasure with a detector has never been easier than it is today. This is especially true with today's brand-new instruments that offer *Imaging by Garrett*. These marvelous detectors tell you the size of a detected target *before* you dig it.

Almost daily I hear stories of treasure hunting successes that proclaim marvels of metal detectors, both old and new. Some of these stories are retold in this book.

The hobby is a grand and glorious one for so many whom I've known over the past half-century. I can only wish that you will find the same joy that they — and I — have experienced. Read on to learn all about it!

And, maybe someday...

I'll see you in the field!

Charles Garrett

Garland, Texas
Autumn 1998

Belgian Finds Everything, But Mostly Old Coins

*S*ome days you dig a centuries-old thimble; another day you dig up a World War II Sherman tank. This could well be the motto of Franco Berlingieri, Belgian metal detector hobbyist. He has a collection of ancient thimbles, and, yes, he once detected a buried U.S. Army tank.

Yet, coins are his major interest and his finds have literally spanned the entire breadth of human endeavor from modern coins back to those used in the Celtic era before the time of Christ. Franco cares little for contemporary coins (except to spend them!); his emphasis is on rare Roman coinage.

"I have proved to myself how effective Garrett detectors can be in finding treasure. My challenge is to use these detectors properly and to hunt with them in the right places. The most "right" place his research ever took him was a site where he discovered 350 Roman coins. Franco lives in Belgium, but takes his GTA along to the Mediterranean island of Sardinia.

Coins are where you find them, and Franco Berlingieri of Belgium seems to find them everywhere! Last year he found upwards of 8,000 and sees no reason why he can't exceed this number in future years.

"What a wonderful hobby this is...treasure hunting with a metal detector."

Over

Various types of quality detectors are available, and this instrument with microprocessor controls was carefully chosen to fulfill specific tasks.

Facing

Coins in the park are the first treasures usually sought, and this uncomplicated OneTouch metal detector is unsurpassed at finding them

Treasure Hunting

*B*elieve it or not, my first "buried treasure" wasn't one that I found but one that I hid…at age four. I had assembled what I considered a marvelous collection of clothing buttons from ladies' dresses. I truly valued this treasure and, wanting to safeguard it, I buried it.

And, do you know what? As far as I know, it's still buried there in Lufkin, Texas. I know that I never dug it up. That personal real life story taught me something. People *bury treasure*, and sometimes they never recover it. That's

Over

Finding treasure with a detector is profitable and fun, and can also lead you into beautiful scenery such as shown here in the Rocky Mountains.

Facing

This prospector used his metal detector to find traces of gold which he is scraping from the mountainside so that he can pan the material.

what treasure hunting is all about...finding valuable items that other people have left behind for one reason or another.

Wouldn't you like to find a hidden treasure? Sure you would! This is something that everyone occasionally dreams about — whether they're willing to admit it or not. Each of us continually seeks wealth in one way or another...and we usually realize results from these efforts as we strive for success. In fact, I believe most of us get just about what we deserve. But, wouldn't it be nice if we occasionally received a "bonus"...the surprise of a hidden treasure.

Yes, the appeal of this dream is universal, and it translates into the rapidly growing attraction of hunting with a metal detector. The desire to seek and find hidden wealth is as old as mankind. This activity transcends all boundaries of age, sex, personality and social status. Furthermore, the hobby of metal detecting offers opportunities that can be matched by no other...even as it gives each individual the chance to fulfill the ancient dream of finding hidden wealth.

I want to convince you that this hobby is one that can literally improve the way you live. The *treasure* mentioned in its title might turn out to be something quite valuable found by your detector. Even more importantly, however, the treasure could prove to be a genuine enrichment of your life made possible by this new pastime...no matter what you might find. Just remember that metal detecting is a delightful hobby that often will pay for itself!

Hunting for treasure with a metal detector is difficult to define because it can mean so many different things. Basically, it is the search for and recovery of anything that *you* consider valuable. That's right, only you can determine what's treasure and what's not. A child's favorite plaything, your lost class ring and a missing house key or earring are treasures just the same as coins, gold and jewels or whatever else you might be looking for.

No matter what the target, metal detecting is an absolutely universal outdoor hobby! Anyone can hunt for treasure with equal intensity anywhere on the face of the earth or under its waters. Each individual decides how much energy is required for participation in the hobby, and the decision can be changed from day to day or even from one minute to the next. The hobbyist can hunt for hours a day or for just a short while; the hunting may be strenuous or involve little exertion.

Treasure hunting sites are equally optional. While hidden wealth can be sought at exotic foreign locations, many successful hobbyists swear that the ideal hunting ground is one's own back yard or neighborhood...the areas that *you know about*!

Some people want (or, expect) to strike it rich the very first time they turn on a detector, while others remain content to find little more than a few coins in the local park. Some individuals hunt with a detector for the potential excitement of digging treasure from the ground, while others are fascinated by the "historical" discoveries they make. Some delight in finding lost class rings in the surf and returning them to their owners. Others enjoy displaying in their homes the treasures they have searched out. Yet, there are many hobbyists who simply appreciate another opportunity for getting exercise in God's great outdoors. Any treasure they find is icing on the cake.

A Universal Hobby

Treasure hunting is an ideal hobby for *anyone*. It channels the natural energy and curiosity of young people, while providing opportunities for essentially harmless adventure and excitement. The hobby is equally as suitable for mature men and women — yes, senior citizens — whose health permits (or requires) light outdoor exercise. Important to older men and women can be the aura of adventure this hobby will bring to otherwise placid (purposeless, perhaps?) lives. Treasure hunting generates an opportunity for safe, yet genuine, excitement and suspense without requiring the rigors or expense of lengthy travel and elaborate equipment.

Hunting with a metal detector has become a family hobby...husband, wife and children can all be dedicated treasure hunters. Sometimes only the man of the household enjoys the hobby, but I know of some wives who will go hunting alone when the husband can't go. Often, children become more proficient than their parents or grandparents. Hunters, fishermen, campers, vacationers and backpackers are adding metal detectors to their sports gear.

Truly, seeking lost treasure fascinates everyone. And, it is a hobby that can literally pay for itself since it offers potential monetary rewards as well as the benefits of healthy exercise and outdoor activity. Finally, no matter what a person's age, health, financial status or social standing...nothing can compare with the sheer thrill of discovery — whether it be that first coin...a ring...a gold nugget...an outlaw cache. The joy and excitement enrich both the spirit and the pocketbook.

Hunting with a metal detector is healthful!

The hobby of hunting with a metal detector obviously takes you outside into the fresh air and sunshine. Scanning a detector over the ground, stooping to dig targets, hiking into new areas...all of this can stimulate the heart and lungs while putting demands on unused muscles. Yes, here we have an extra side benefit. A "built-in" body maintenance program is a valuable *plus* of treasure hunting. Leg muscles stay firm, flab around the middle diminishes, breathing improves and nights of restful sleep foretell a longer, healthier life.

Most important to many people is the awareness and enjoyment of treasures of nature that have been placed upon the earth for all of us to find. Whether you discover items of monetary value is never really the point. What is always truly gratifying is to see nature in its purest form all around and to be a vital part of it. This alone could well be your greatest treasure discovery.

Educational factors related to treasure hunting can be equally stimulating. Relics and artifacts of bygone eras raise many questions. What happened to the people who lived and prospered where only faint traces remain to mark the existence of a once-thriving city or town? Why was the area left deserted? These and other mysteries can sometimes be solved with proper research and examination of the artifacts you discover.

For the hobbyist who takes pride in his or her efforts the greatest difficulty is deciding which of the dozens or hundreds of leads to follow. The common sense that comes with age and experience is a big factor in metal detecting just as it is in any other human activity.

Many successful hobbyists seem to live and breathe the hobby. It is always on their minds. Consequently, everything they read and everyone they talk with are potential sources of the fresh treasure leads and data needed to help in their work.

To find treasure you must prepare yourself with the right attitude. You must remove obstacles to clear thinking by developing faith that treasure is *there* to be found and that it *can* be found. Then, if you will diligently apply yourself, use the proper equipment and search according to the rules set forth in this book, your chances of locating wealth become overwhelmingly great.

Sure, luck plays a part in treasure hunting, just as it does in any other aspect of life. But, I agree with the great baseball executive Branch Rickey who defined luck as "the residue of good planning." Hard work and using the proper equipment will always pay off. Never forget that.

So, first, get a good detector and learn all about it — its searchcoils, modes of operation and capabilities. Then, conduct a little research...even, if the only source you use is your own memory. Finally, put your study into action with practice, especially in the field. How hard you hunt, what hours you keep...all of this is strictly up to you. But, even if you hunt only occasionally, you can enjoy success with a modern OneTouch metal detector. They're so easy to use!

At the opening of this chapter I stated my belief that most people get just about what they deserve. Well, you know what? The same is true of metal detecting hobbyists. Those who learn to use their detectors and conduct research properly generally find the most treasure. The prizes they find are not "bonuses"...indeed, they are the well-earned results of carefully planned efforts.

If you will equip yourself with a good detector, learn how to use it properly and base your hunting on just a little research, you too can be successful. I sincerely hope that such will be the case with every person who reads this book.

Once you get that detector and become familiar with the few "basics" explained in the next chapter, you can go treasure hunting yourself with no further ado!

Detector Leads Salvors To Sunken Treasure

*M*el Fisher had been searching for the fabled Spanish treasure ship *La Señora de la Atocha* for more than fifteen years. His quest had cost a small fortune as well as the lives of his eldest son and daughter-in-law. The search had reached into the *Archives de las Indias* in Seville, Spain, as well as courtrooms over the United States.

One fact was indisputable: the Atocha had sunk in a hurricane in 1622 laden with a fortune in gold, silver, jewelry and emeralds. But, where? That question that kept driving Mel and his faithful followers.

On Saturday, July 20, 1985, Mel's salvage ship *Dauntless* was anchored 41 miles west of Key West. Two divers with a Garrett Sea Hunter metal detector were exploring the ocean floor when one spotted silver coins scattered about the ocean bottom. As he fanned sand from these coins, the other diver swam with the Sea Hunter toward what appeared to be a large mound of coral. When he casually scanned his detector over this "coral," it sang out so loudly and insistently that he almost dropped it. The coral-covered mound turned out to be nearly solid silver…hundreds of silver bars and thousands of coins.

Mel Fisher had finally located the mother lode of the Atocha, the largest sunken treasure ever found.

Basics Come First

*I*f you want to begin hunting right now, this chapter will answer enough of your questions to get you started. If you aren't interested in using a metal detector right away or this chapter seems too "technical," just glance over it and continue with Chapters 3 through 7 for a better understanding of all the rewards that can come from treasure hunting.

Now, to answer commonly asked questions...

Q. Is there something in particular I should do before I begin metal detecting?

A. Yes, read the instruction manual that accompanies your detector and watch any video that comes with it. Actually, you should do this even before you assemble your detector.

Q. Where should I hunt?

A. Where you believe coins or other metallic items of value might have been lost. And, wherever people have been, they probably lost coins and other treasures. This can be on private property where you have the owner's permission or on public property where the use of metal detectors is permitted. Chapters 9 and 10 get quite specific on this subject.

Q. After I turn on the detector, what controls do I set?

A. None…at least, with a modern OneTouch detector. They've all been set at the factory for you. You're ready to begin scanning. Walk slowly and methodically in a straight line and sweep your searchcoil back and forth in front of you. Let the searchcoil skim along at a height of one to two inches.

Q. How will I know when I find something?

A. Your detector will sound off when it detects metal. You can look at the target identification (LED or meter) to get a better indication of what you might have found.

Q. How deep will my detector find targets?

A. You can expect a modern, computerized instrument to find coins at depths sometimes exceeding nine inches and larger targets even deeper. As explained in Chapter 12, however, the only honest answer to this questions is, "It depends."

Q. What's "Discrimination" all about anyway?

A. The circuitry of modern detectors enables you to choose which targets you want to find. Factory settings of most detectors will include some discrimination when you turn them on, but you can regulate this to find all metal targets or to eliminate any that you choose. With a modern detector you won't need to be concerned with discrimination just now, but you'll want to learn more about it from reading this book, especially in Chapter 12. The instruction manual or video that accompanied your detector will explain specifically how it utilizes discrimination to make treasure hunting easier for you.

Q. Should I wear headphones when hunting?

A. They can be quite helpful. You hear signals better and can often hear more of them, especially in areas where there is loud ambient noise from people, traffic, surf or running water, etc. Plus, you really don't want bystanders to hear the sounds your detector is making.

Q. I guess there's a lot more I should learn about?

A. Yes, to say the least; but this will get you started. Remember, however, that even after you begin working with your detector you'll want to read on in this and other books to learn about treasure hunting so that you can discover more than your share of "found" treasure. This will come only after you understand and develop proper techniques. Throughout your treasure hunting career, however, I urge you to remember to be patient and to persevere. Good Hunting!

And, did you know that 95% of all detector hobbyists search for coins? The next chapter will tell you why.

Hobbyist Discovers
Priceless Coin

*T*om Brown is a veteran metal detector hobbyist who believes in research and in being careful with anything he finds. And finding is certainly the name of the game for this New Englander.

Since "moving up" to a Master Hunter CX model he's recovered large cents, a 2-cent piece, Indian Head cents, a Civil War token and an 1818 Spanish-American reale.

"I concentrate my effort in old towns and use detective work to find once-popular meeting sites, such as churches. When I find an open field, I head for the oldest tree and begin looking there."

Tom knew that he had found something old that day in the sandy soil of southern New Hampshire near the Merrimack River. Responding to a signal from his CX, he unearthed a silver coin that was literally indescribable. But, Tom handled it with care.

When he removed the dirt and saw the date 1652, he knew he'd found something valuable. What he'd found was a bit of American history, one of the earliest Colonial coins. It was a Pine Tree threepence minted by members of the Massachusetts Bay Colony in defiance of the British crown. It's value? Priceless, of course, but valued at thousands of dollars by collectors.

3 ~ Can I Really Find Them?

Lost Coins

*W*hen you start out with a modern OneTouch computerized instruments, especially one with imaging, you'll soon be hunting like an expert. With surprisingly little effort you'll even find yourself discovering coins that other hobbyists have passed over! And, imaging will help you avoild digging those tin cans and other trash targets that slow down the hobby..

When most people think of hunting with a metal detector, they visualize someone searching for pennies that children have lost at a playground or in the park. And, they have the right idea because this is an important part of the hobby; such hunting can be quick, easy and *profitable*. And, it's the type of hunting that can take place anytime with as many or as few hours spent as the hobbyist desires. Plus, with a OneTouch detector it can be accomplished successfully with very little practice.

But..."Pennies?" you may ask. How can it ever be "profitable" to search for pennies?

Who's talking about pennies? We're going to talk about finding *coins,* and it doesn't matter who lost them — or when!

Hunting for coins is the heart and soul of the metal detecting hobby because literally everybody hunts for coins.

They are certainly the initial target sought by most first-time detector owners of any age. And, why do so many men and women hunt for coins?

Because...they are *there!*

Think of all the coins that you have handled in your life...thousands and thousands...and, think of those you have lost. Your experiences were shared by most other folks. So, where today are the millions of coins that have passed through your hands and those of other individuals?

These coins were made of metal. Certainly, not all were used enough to "wear out." Something must have happened to them!

So, think of all of these coins — now "lost" ... and lying atop or beneath the ground, hidden in houses and buildings or in waters of lakes, streams and oceans...coins just awaiting discovery — "finders keepers" coins that belong to the first person to find them or dig them up. Why, the sheer number of these "lost" coins far exceeds and *far surpasses in face value alone* all the coins currently in commerce, savings and collections. And, you know how truly valuable old coins can be.

Lost coins can be found all over the world. Except perhaps for the polar icecaps and barren Asian peaks, I sincerely believe there is no place on the face of the earth where coins cannot be found with a metal detector...and,

Californian Ed Morris spends only a few hours a day treasure hunting but his array of coins and other finds is impressive.

in abundance, if the site has never before been searched. Literally multitudes of coins can be found everywhere. You just have to look for them. Why am I so sure about this? I've proved it for myself!

Coin-hunting is the usual introduction to the fascinating and rewarding hobby of searching with a metal detector. Let me warn you, however, that you'll soon find yourself "hooked." After you begin finding coins — and especially when you make that first "big" find — coin-hunting will be ever on your mind. No matter where you are or what you are doing, you'll find yourself subconsciously evaluating the coin-hunting prospects of that location.

And, you know what? *You'll love it!*

If you are totally new to the hobby of metal detecting, you may already be questioning your competence as a coin hunter. And, honestly, aren't you a little suspicious that all you'll get from this hobby is exercise and fresh air — which you probably need, anyway. But, I hope you're already convinced of how easy it is to find coins with a modern metal detector. And, if you need suggestions about places to find them, you'll find plenty in Chapters 9 and 10.

Where Do I Look?

First of all, study Chapter 9. Then, remember that we must always search for coins *where* they have been lost, and we must always search with the *proper* equipment.

In recovering the coin he has detected, this hobbyist is being careful not to scratch it and will thoroughly inspect the hole he dug before filling it properly.

But, where to look is actually easy! Anywhere people have been — which is practically everywhere. Once you get really interested in this hobby, you'll soon have the problem of so many places to search that you truly won't know where to go next.

I find it impossible to stress adequately the importance of research. It's truly the *key to success* in finding treasure with a metal detector. Certainly you are glad to pocket the "profits" of finding coins anywhere. You will discover, however, that the greatest personal rewards of this hobby as well as those of most monetary value come from finding valuable old coins through your own research, investigation and hard work. Even when you come home with a full treasure pouch, you may grow tired sometimes of digging up current coins in the parks and playgrounds. You'll never grow tired of recovering old, rare and valuable coins in places *you* discovered by your own desire and careful investigative efforts. You will only become more enthusiastic and your rewards will increase.

You may remember history from school as a jumble of dates and names, all dry as dust. When you're looking at a history of people who lost coins and valuables in places you can search, however, you'll find that it takes on a new and more attractive life. To me research can be as much fun as the actual treasure hunt. And, it's a part of this hobby that you can pursue any month of the year in any kind of weather. Plus, a good researcher will never find the time to follow through on all the great tips that he can discover.

I know because, unfortunately, I'm speaking from the experience of having far more places to search than time to search them. Always remember to research your own memory. I had often thought about people losing coins at the Harmony Hill "lovers' lane" that was so familiar to young people growing up in my hometown of Lufkin during World War II. My brother and I finally searched it recently with metal detectors. I smile when I think of the silver coins we dug there. And, we'll probably find more when we return for a final search!

I believe that every metal detector I've ever designed and built was used at one time or another to search for coins. I'm also confident that many of these instruments were and are still being used for no other purpose. Oh, jewelry and other valuable items are certainly pocketed when found. And, for one reason are another, they are occasionally sought. Coins, however, are the thrust of the search for most hobbyists, no matter what their intentions. So, all detector hobbyists should master coin-hunting techniques which will be used in nearly all other forms of treasure hunting.

A computerized OneTouch detector is ideal for finding coins anywhere. In fact, most of these detectors will be used by coin hunters. The precise target identification possible with microprocessor-controlled circuitry permits detection at extreme depths. Combine this with imaging that lets you know the size of targets and a system of graphic target identification with an ID scale, plus ease of handling, and you have a superb instrument for finding coins.

Whenever you turn on a computerized detector, it's automatically ready to hunt for coins...whether you're in a park or on a beach or wherever. Circuitry is preset at the factory to respond ideally to conditions normally encountered. Its discrimination is designed to eliminate detection of lower conductivity trash targets normally encountered in coin hunting, such as bottlecaps and most pulltabs.

Coins & Junk

There will be times when you hear a good solid detection tone while scanning over a spot in one direction and just an audio "blip" when scanning from the opposite direction. Although this is most likely a junk target, we advise you not just to walk away from it. Continue scanning back and forth across it from numerous angles. Draw your imaginary "X;" slide your searchcoil from side to side. Push the coil forward and backward. Listen to the sound carefully and check your graphic display. If you ever get the coin tone in both directions (headphones are a real help here), dig the target, even if it doesn't register as a coin on the graphic display. You'll probably find a coin in close proximity to some sort of junk.

We know how much willpower it takes to resist these blips coming over your headphones or from the audio speaker of your detector. You know some sort of metal target exists, and it's human nature to want to dig them all! Remember, however, that a quality modern detector will never lie to you. Its graphic display and audible sounds will generally identify all objects being detected beneath the searchcoil at any given instant. My friend Ed Morris of Santa Maria, California, relies on the graphic target ID display of his detector to such an extent that he refuses to dig when it indicates that he has found only a penny.

Of course, because adjacent trash can sometimes "confuse" the detector, you may want to dig all targets. That's what I often do...especially where there's money with a lot of trash around it!

I recommend that coin hunters begin with the standard 8- or 9-inch general purpose searchcoil for scanning most parks, playgrounds, beaches and other conventional coin hunting areas. Because it is the best all-around searchcoil, such a model is usually standard equipment with a computerized detector. You will really never need to use another coil in your enjoyment of the hobby.

Numerous letters and personal inquiries, however, regularly ask about the occasional use of larger or smaller coils. My answer is a definite *yes*...because larger coils can detect deeper, yet find even the smallest coins, and smaller coils also offer special advantages.

There is a misconception that large coils cannot detect small objects, but this is not true with today's detectors and coils. I have used our 12 1/2-inch model to find tiny objects at great depths. Of course, you'll need experience to realize the benefits of using a larger coil to find coins because while the bigger coils will find more deep targets, they'll also find more junk. And, you should use headphones because signals from deep targets may be weak. Because of these and other factors, I recommend that you limit your use of large coils in the beginning to cache hunting and searching for veins of precious metal.

But, after you scan an area with a general purpose coil and are certain that you are encountering deep targets that give you only a faint signal, even with headphones, you can get additional detection depth with a larger coil. After scanning such an area thoroughly with your general purpose searchcoil, go back over it again with a larger coil, scanning very slowly and using headphones. If there are deeper coins to be found, you'll detect them!

Smaller coils inspect a *smaller* area and increases your ability to zero in on the coins when there's an abundance of junk targets around in the ground. Used properly, smaller coils will make you more efficient in your coin hunting efforts. With the smaller coil helping you concentrate on good targets, you will dig a higher ratio of coins to trash, even though detection depth will not be as great as with larger sizes. Smaller coils should definitely be weapons in your normal coin hunting arsenal. You can handle them a lot easier too, and they will "fit" in more places than a larger searchcoil and let you scan closer to metal poles, building foundations and such.

Recovering Your Finds

You're going to hear a lot of advice from veteran hobbyists about digging coins:

"Don't use a screwdriver...it'll scratch!"

"Never use a knife...you'll ruin the sod!"

"Always use a probe...less damage to the sod!"

"Don't scratch your coin with a probe!"

Sounds confusing, doesn't it? Well, let me confuse you just a little more by telling you that all of the above advice can be true...it just depends on what you're digging for and what kind of turf you're digging into! The problem coin hunters face when it comes time to dig is that each type of soil requires, generally, a different digging and retrieval technique. Retrieving coins from the sandy beach, of course, is perhaps the easiest. Digging coins out of hard-packed loam under dense St. Augustine grass may be the most difficult. Retrieval in loose, sandy soil under a growth of Bermuda grass lies somewhere between the two extremes.

My coin-hunting expeditions have taken me from frozen turf in the Far North to sandy Caribbean beaches, and I've stopped at lots of places in between. So, I understand a great many of the different techniques. But, they must be learned by any coin hunter who wants to hunt with maximum efficiency. I wish I could tell you that somewhere in these journeys I had discovered a magic solution for digging. But, there is none. It just represents work...sometimes, harder than at other times...but, always work. The only consolation is that there is often a fine and immediate reward for the work!

In loose and dry sand it's quite easy to use a wire mesh scoop-sifter and a small plastic trowel something like a sugar scoop. When you find a coin under well-kept sod, you can have a problem. One solution is to use a sharp knife and cut a half-circle around your discovery; make it three inches deep. Then, fold the turf back. If the coin isn't in this first plug, remove a second and deeper plug, making certain that all loose dirt falls back into the hole. After your retrieve your coin, replace all the soil you've removed before folding the turf back in place and stepping on it. Of course, you should never put trash back in holes for someone else to dig up. That's why you should wear an apron with two pockets...one for keepers and one for trash.

Some veteran hobbyists recommend that you cut your plugs square so that they will fit precisely back into the hole. They will also suggest that the plug be cut deeply with lots of soil attached to it so that powerful lawnmowers can't pull your plug from the ground before the grass has become re-established.

Of course, a screwdriver is one of the most popular tools used today for coin retrieval. To use one push it into the ground about two to three inches behind the coin. Stick it in at a 45-degree angle about five inches deep. Of course, you'll want your screwdriver to have a dull point to keep from scratching a coin if you should accidentally punch the point into this find. Now, with your screwdriver inserted five inches push forward and to one side, making a slit in the ground three to five inches long. Then make the same slit to the other side, with the two slits leaving a "vee-shaped" piece of sod. Push the vee-shaped chunk of sod forward with your hand, swinging it up out of the ground. After you have retrieved your coin, the sod will fall back into the hole in the exact place it came out.

Here's a slightly different retrieval method for use in parks and other areas where the hobbyist must be careful not to damage the sod. After pinpointing, carefully insert your dull-pointed probe into the ground until you touch the coin. This will inform you of its exact depth. Then insert a heavy duty screwdriver in the hole made by your probe, but stop before it touches the coin. (Remember, you already know how deeply buried it is!) Rotate the screwdriver gently until you have a cone-shaped hole about three inches in width across the top. It is then usually an easy matter to remove the coin with just a little digging with your fingers or the point of the screwdriver. This method requires some practice and skill, especially when probing, because the coin must not be scratched. To fill the hole, insert the screwdriver into the ground two or three times around the opening. With just a small pressure toward the hole the surrounding soil and grass fill it in, leaving absolutely no scar.

This whole matter of recovering coins is one that hobbyists can spend hours discussing. My popular book, *New Successful Coin Hunting*, devotes an entire chapter to the subject. We're always ready to learn new ideas!

Good luck with your digging, and always remember to *fill those holes*!

On Your Travels

You'd be amazed at the places where you can hunt coins just in your normal weekend and vacation traveling. For example, how about those old drive-in theaters you see along the roadsides? There aren't too many left, and the sites are fast filling up with houses and shopping centers. Still, there must be many coins in these fields with numismatic value equivalent to the cost of our finest metal detector. Deserted highway rest stops and cafes, roadside parks, camping, hunting and fishing parks can be found along many roads. Scanning your detector at locations such as these can not only prove profitable, but can provide an opportunity for stretching your legs, walking the dog or making new friends for our hobby.

When you get off the Interstate Highways in your travels, you pass through many smaller towns and communities. Most of these have some kind of park, playground or swimming area. Drive to the parks and let your wife watch the children play while you search the most likely places for rare coins. Don't forget to fill all holes you dig and to dispose of any trash you encounter or dig up. This may help calm that caretaker who wondered what you were doing digging up his park. Always visit historical markers. Many travelers stop here, rest, take pictures and lose coins. Who knows? You might dig up a treasure cache like those which have been reported found around prominent historical and state border markers. And, be sure to read all the historical plaques and monuments. You may get clues to nearby locations or the actual sites where relics and other treasures are located. Plus, you'll learn some more history!

Off the Beaten Path

When you have time in your travels, get off the main roads...especially if you're serious about coin hunting. The farther you can get from today's civilization the more likely you are to find old settlements and places where coin hunting is good. Why not check out the map before you leave on your next driving trip? Look for alternate routes. Take the back roads, and drive a little slower.

There are literally thousands of obvious places that may surprise you with the quantity of good old coins they yield. I'm talking about courthouses, parks, community recreation centers and such public places that have *never* been searched. You may be the first person ever to scan a metal detector over them. But, if you look for leads, you can do even better. Talk to people who are familiar with these little towns...people who can direct you to old campgrounds, settlers' meeting places, old fairgrounds and peddlers' stands. There are always some old timers sitting around the courthouse square or on the benches that remain on small town streets.

One approach that I've laughed about over the years is to go up to these senior citizens and ask, "Say, can you tell me where I can find an old timer in this town?" That generally brings a laugh and helps open them up. When they start talking about the past and where people gathered, you'll be amazed with the volume of information you can compile and how quickly you can gather it.

One story I like concerns an ethnic reunion that is still held annually in a Central Texas area. Literally hundreds of men, women and children gather once a year in this particular location for a rousing picnic that seems to be climaxed each year by considerable drinking of beer and wine and more than a few friendly scuffles. Obviously, the location for this annual get-together is filled with coins...and the site is replenished annually. Yet, this is the type of information that can be found only through personal research with local citizens.

Did you make a wrong turn on that unmarked road? What's your hurry? Just drive along and see some new scenery until you return to your proper route. Stop and ask directions. Maybe you'll discover a new coin hunting location that you would never have found by taking the "right" road.

Picnic areas and roadside parks are favorite places for the traveling coin hunter. Search carefully around all tables and benches and out away from seating, especially in grassy areas. Look carefully around drinking fountains, along trails and around trees. Frequently, after eating, travelers will wander into grassy areas and lie down to rest. If areas are large, search places that are shady. Search where cars have been parked.

Churches and Brush Arbors

One of the oldest nickels I ever recovered was found five feet straight out from the front doorstep of an old church. Five inches deep, it gave an excellent signal in our mineral-free East Texas soil. When searching around churches and tabernacles, look especially in front where people might have stood and talked after the services. They lost coins here. Also look around back near trees where children might have played. Search areas where cars and buggies would have been parked. Old churches also usually had picnic areas which you should try to locate and search.

Never pass up the chance to search any area that was ever used as a "brush arbor" or site for outdoor revivals. A scattering of sawdust, long benches and an overhead brush covering provided an instant worship location. You can be certain that coins were lost here.

As an old friend liked to recollect, "Can't you just imagine some sleepy fellow sitting on the hard benches trying to stay awake and pay attention to a sermon. He gradually drifts off to sleep and, first thing he knows, someone is punching him in the ribs and waving a collection plate in his face. Of course, he's embarrassed and fumbles out some coins to drop in the plate. He *also* drops some coins in the sawdust below where they are still waiting to be found!"

On the Beach

No chapter on coin hunting would be complete without some discussion of searching for coins and other lost wealth on the beach. It's truly the new frontier for treasure hunting, and it's one that is rich with rewards. Any public beach, whether of an ocean, lake or river, is a good place to try your luck. More on this in Chapter 5!

Let me repeat that this chapter on coin hunting is designed to stir your imagination...to help convince you that coins can be found everywhere. I have proved time and again that this is true, and so have millions of other hobbyists. Only by getting out in the field yourself — with a quality metal detector, preferably — can you appreciate the true excitement and the real joy of this hobby...excitement and joy, incidentally, that are substantially heightened by the *profit* motive!

And, speaking of profit...what could be more profitable than real gold? Read on to learn exactly how *you* can use a modern metal detector to find gold nuggets.

Gold Nuggets Are Where You Find Them

*O*kay, so it was Georgia and not the Rockies or Sierra Nevadas. But, from studying old books and newspapers he knew that gold had been found here in the past. Standing in waist-deep water, Bill Boye swept the searchcoil of his Garrett detector over bedrock and received a loud signal. "Just another tin can or worn-out tool, I guess."

He used a crowbar to break a large piece of quartz off the bottom and checked the hole again. Surprise...no signal! As Bill swept his coil across the chunk of quartz he had dislodged, he was startled both by the egg yolk that appeared to be smeared across one side of it and the loud signal it caused his detector to make.

"Wait a minute, egg yolk would have washed off? And, where would it have come from?"

Bill shook his head to help clear his mind, and his hands began to shake as he realized what it had found. Measuring 16 1/2 by 6 inches and weighing almost 17 1/2 pounds, it's the largest single gold specimen found in North Georgia since the gold rush there two centuries ago.

Sure, there's no gold to be found east of the Rockies, and certainly no large nuggets. But, don't tell Bill Boye whose research and perseverance paid off.

Gold Nuggets

*H*unting for and finding gold with his (or her) instrument is something that just about every detector hobbyist *dreams* about. I believe this to be true...even if the hobbyist hunts almost always in a nearby park and really has no plans ever to try anywhere else.

Well, what's wrong with that? A guy can dream, can't he? And, dreams are a great deal of what this hobby is all about!

So, this chapter that explains hunting for gold with a detector will at least help you dream more realistically. And, if fate (or, a vacation) should take you into the gold fields, you'll know how to use your detector there.

Can I really find gold with a metal detector?

This is the question that I hear most often from gold seekers who doubt their ability — or *anyone's* ability, for that matter — to find gold with a detector.

The answer to that question is a most emphatic *Yes!* I have proved, many times over, the abilities of metal detectors for finding gold.

Gold Panning Instructions

1. *Place the classifier atop the large gold pan and fill with sand and gravel shoveled from bedrock.*

2. *Submerge the classifier contents under water and use a firm, twisting motion to loosen material. Gold, sand and small gravel will pass through the classifier into your gold pan. Check for nuggets in the classifier and watch for mud or clay balls that might contain nuggets.*

3. *Discard all material remaining in classifier. Use your hands to thoroughly loosen all material in the gold pan. Inspect contents and remove pebbles. With the pan completely submerged twist it with a rotating motion to permit the heavy gold to sink to the bottom.*

4. *Keep the contents submerged or covered with water. Continue the rotating, shaking motion. From time to time tip the pan forward to permit water to carry off lighter material. Pour it so that material passes over the riffles.*

5. *As you shake the pan to agitate the contents make certain that it remains completely submerged and that the Gravity Trap riffles are on the downside. The lighter material will float over the pans edge while the riffles trap the heavier gold. Rake off larger material from time to time.*

2

3

4

5

6

7

8

9

10

6. *Develop a method of agitation with which you are comfortable. Back and forth...round and round...or whatever suits you. Your aim is to settle and retain the heavy gold while letting the lighter material wash across the riffles and out of the pan. As the pan's contents grow smaller, smooth and gentle motions are mandatory. Use extreme care in pouring lighter material over the side. Submerge pan often and tilt it backward to let water return all material to the bottom of the pan.*

7. *If there is a larger than usual concentration of black sand, you may wish to transfer the material to the smaller finishing pan for speedier separation.*

8. *Continue the panning motion to let all remaining lighter material flow off the edge.*

9. *Now, you can retain the black sand concentrates or continue gentle motions to let it ease of the edge of the pan. As visual identification becomes possible, a gentle swirling motion will leave your gold concentrated together.*

10. *Retrieve your gold. Use tweezers for the larger pieces and the suction bottle to vacuum fine gold from the small amount of water you permitted to remain in he pan. Save the remaining black sand for later milling and further classification at home.*

I also want to stress to you the pleasure (and, profit) that can come from panning for gold. It's a lot easier than you may imagine...both the mechanics of panning and your ability to find *some* gold. As a veteran amateur prospector in Utah told me recently, "It's easy to find gold. It's just hard to make a living at it!" What this means to you is that you can find traces of gold quite easily...if you're willing to work for it.

A special six-page section in this chapter explains clearly and shows in photographs how you can pan for gold and bring back a unique souvenir to help you remember that vacation out West. Garrett offers a complete kit that includes everything needed for gold panning. Use the order blank at the end of this book for this kit or for books on hunting for gold.

Nugget Hunting

I can assure anyone who follows just *three* basic rules that he or she can be virtually certain of finding at least some gold or other precious metal with a metal detector.

Rule 1 — Choose the correct *type* of detector.

Rule 2 — With *patience* and *perseverance* use the same scanning techniques you learned in hunting for coins.

Rule 3 — Hunt in areas where gold has *already proved its presence*.

These rules come from my close friend, veteran prospector Roy Lagal, who has been successful in finding gold with a metal detector for many more years than the 25+ we have known each other. Most of this chapter represents his wisdom and experience.

Let me emphasize that hunting for gold with a metal detector is no task for the absolute beginner. Before you take a detector into the gold fields, please use it for many hours hunting for coins, jewelry and easier-to-find targets. You'll be glad you did because tiny gold nuggets are far more elusive than even the smallest coin. Believe me!

Before we learn how to find gold, however, let me pass on a piece of advice. No, this is more than advice; it's a strong suggestion. Make certain that before you buy a "gold-finding" detector that the instrument is also equipped to find coins, jewelry and similar items. Some aren't! Remember that all hobbyists may dream about hunting for gold, most of them generally spend a lot more time seeking coins, jewelry and the like.

Rule Number One, probably leads to two questions: What is the *correct* type of detector? Well, "correct" doesn't necessarily mean some particular brand or model...although Roy obviously favors the Garrett instruments which he has helped me to develop over the years. And, second, yes, you can use a general purpose computerized detector to search successfully for gold nuggets. When you get really serious about hunting for gold, you can then consider a model expressly designed to find gold.

Gold-hunting models offered by the various manufacturers have proved themselves time and again all over the world. Yet, some of these same detectors offer a hunting mode that enables them to find coins with ease. With one of these detectors hobbyists can find coins when they tire of hunting for gold. (Or, when they find themselves in an area where there's just no gold to find!)

Modern computerized detectors with their precise circuitry that features high sensitivity can completely balance out *all* effects of negative mineralization. Hunting for nuggets in gold country can be just as effortless as searching a park for coins...as far as the effects of ground searching a park for coins...as far as the effects of ground mineralization are concerned. Gold nuggets can be found amid mineralized rocks with ease.

Patience

You must have it. Learn to understand your detector fully and become proficient in its use. Take your time scanning in the field and don't get in a hurry. Try not to get discouraged when results are disappointing. And, if all else fails, fall back on this tested prayer:

Lord, please give me patience, but be quick about it because I don't have time to wait!

When you're hunting for nuggets, use absolutely no discrimination and examine all targets. Those of you who know even a little about the hobby understand just how much patience these two instructions will require. You'll discard a lot of trash and a lot of rocks that "look like" gold. And, if you bring some of them home, don't be embarrassed. Even the best of the old miners often had their hopes dashed with "fool's gold."

You'll notice that our second rule also includes the word *perseverance*, which the dictionary defines as persisting "in spite of discouragement." That's a good definition for gold-hunting because it can sometimes be a discouraging business...hour after hour after day after day.

Rule Three concerns where you must start your search. Some research is a must here. No one can find gold or any other precious metal where they simply do not exist. Confine your searching to areas that are known to have produced gold until you have become very familiar with the telltale signs of mineral zones. And, if you should decide to strike out on your own into an untested area, plan to rely even more heavily on Rule Two!

Yet, there are so many things that are now possible with modern electronic metal detectors! An entire vista has been opened up by truly dramatic technological improvements. Totally new areas of opportunity are being revealed to even the most veteran gold hunters. Novices are fortunate indeed to be able to begin their electronic prospecting careers with the 21st-century detectors that are available today.

If you're interested in finding gold with a metal detector, I suggest that you read the new *You Can Find Gold with a Metal Detector,* which Roy and I just wrote. It contains numerous tips that will let you build on the expertise you developed in coin hunting.

Unbelievable success can await you if you will use the *correct instrument,* conduct research thoroughly and employ the virtues of wisdom, patience and persistence.

And, now that we've talked about finding nature's gold, let's learn how to find gold jewelry...on the beach!

Detector Enthusiast Helps Avert Tragedy

*C*hris Martin's spring break trip to Florida had become a disaster. Before he went into the ocean, he carefully put his new senior ring in his cap. When he rushed from the surf and donned the cap again, he forgot about the ring, and the Dallas youth lost forever (he thought) a new possession he had already come to treasure.

Yet a few days after returning home Chris was amazed to open a mysterious package and find the very ring he had "lost."

Bob Christensen of Palm Bay to the rescue again!

Bob is a self-appointed beach patroller in Central Florida whose tee shirt proclaims "I look for coins for kids." All coins he finds with his GTA 1000 are donated to the Palm Bay Parks recreation programs, as well as rewards he receives for finding lost jewelry or any loose change he can "talk a beachgoer out of."

Plus, Bob is no amateur at finding coins. How about 3,000 or so a year?

Locating Chris was a snap for the dedicated hobbyist. He simply called the manufacturer of the ring, described it and within minutes was talking to Chris' father on the telephone. And, the Texan's generous donation helped Palm Bay children to enjoy that year's annual Easter Egg Safari.

Beach Treasures

*I*n the chapter on coin-hunting I commented that when a great many people think of hunting with a metal detector, they see someone searching at a park or playground. There are others, however, who always visualize a detector being used at the beach. The fact is that beaches may be better hunting grounds generally than parks. The digging is usually easier, and there has always been wealth to be found near the water!

Over two thirds of the earth's total surface — nearly 200 million square miles — is water. Since the dawn of time man has spent most of his hours and days on or near water. Transportation, commerce, recreation, exploration, warfare and the search for food have compelled men and women to return to water time after time...whenever they have strayed. And, whenever man made contact with water, he generally brought along valuable items, some of which were inevitably lost.

The world's oceans, lakes and streams, therefore, offer vast storehouses of lost wealth that await the treasure hunter. Beaches at the entrance to Davy Jones' locker present the most accessible areas for many hobbyists to begin their treasure hunting career.

At the water's edge with today's detectors you'll be looking for all kinds of treasure, ranging from current coins and trinkets to expensive jewelry or relics from shipwrecks. Modern metal detectors are the ideal tool for discovering this wealth. Designed to overcome both the mineralization of beach sands and the effects of salty ocean water, they literally "look" beneath the sand. There was no way — even just a few years ago — that the best instruments could have operated under such conditions.Now, imaging even lets you avoid digging cans and similar junk targets. Yes, today's detectors have opened a new frontier. Come and explore it!

A word concerning equipment is mandatory before beginning a discussion about the beach. Most of the new high quality computerized detectors are satisfactory because their factory-set Beach Modes overcome mineralization and the effect of salt water. In addition the searchcoil for most of today's detectors is fully submersible. You can hunt in water several inches deep, but always remember that the control housing of your detector must *never* be submerged unless you're using an instrument designed for underwater use. Incidentally, the subject of underwater hunting is one that I won't be discussing in this book. If you're interested, Ram offers several books on the subject.

Be careful too about laying your detector down where a wave might wash up. Of course, the instrument should always be protected from blowing sand, rain or splashing surf. The danger of water is obvious, but sand can be the real villain because it tends to seep through the tiniest crack.

My advice to any hobbyist is to become a beachcomber. The joys are countless, and the rewards are constantly surprising! Just what is a beachcomber? I describe him or her simply as a person who searches along shorelines. And, what are they seeking? Just about everything! There's always plenty of flotsam, jetsam and other refuse. Often, it's merely junk, but it can turn out to be lost wealth.

It's hard to understand why people wear valuable jewelry to the beach. Yet, they do, and they often forget...even about precious heirlooms and diamond rings. But, whether sun bathers and swimmers care about their possessions or not, it's the same for the detector hobbyist. All rings expand in the heat; everyone's fingers wrinkle and shrivel, and suntan oils hasten the inevitable losses. Beachgoers play ball, throw frisbees and engage in horseplay. This activity flings rings from fingers and causes clasps on necklaces, bracelets and chains to break. Into the sand drop treasures which quickly sink out of sight.

Always remember that wherever people congregate (or, have congregated), valuables can be found. Beach treasures awaiting the hobbyist include coins, rings, watches, necklaces, chains, bracelets and anklets, religious medallions, toys, knives, cigarette cases and lighters, sunshades, keys and countless other items.

And, this is only atop the sands! Out of sight below lies that real blanket of wealth awaiting the metal detector — the same type of items to be found on the surface, but usually much more valuable because they are older...sometimes centuries older. Consider too that the value of any treasure is ultimately determined only by its finder. Keeper finds can be anything from a weathered float to a costly piece of jewelry. And, for some lucky, persistent and talented hunters, their dream will come true. Even though they may not find that chest of treasure hidden by some buccaneer or

17th-century Spaniard or Frenchman, they will answer the sound of their metal detector by digging a Spanish *reale* from the sand.

Yet, oftentimes, the greatest joy for the hobbyist comes simply from walking the beach, from experiencing soft winds off the water and feeling the sand under bare feet while listening to a detector harmonize with the tranquilizing sounds of surf and seabreeze. Rewards of the hunt are but an added bonus.

When you walk out onto a beach, where do you begin? How do you select the most productive areas? This is possibly the question I am asked most frequently by beginning beach hunters (and, more experienced ones as well). The answer, first of all, is that nobody should go pell-mell onto a beach and begin hunting here and there without a plan. To find treasure *anywhere* you must be at the right place at the right time with the right equipment.

The dedicated hobbyist always first answers the question of "Where?" with research. And, remember, research can mean simply paying close attention to what you see!

Beyond that, experience must be your teacher. Inquiring and attentive hobbyists continually pick up ideas from veterans, but the final decisions must be based on individual perceptions and intuition. Experience can educate the beach hunter about places that never produce and other places that are often rewarding.

Stay alert to current weather conditions. You'll want to search at low tides — the lower the better. After storms come ashore, head for the beach. When oil spills deposit tar and oil on beaches, there's a good possibility bulldozer and other equipment used to remove it can get you closer to real treasure. They'll scrape away the top layer which may contain yesterday's coins, but this heavy equipment might expose deeper treasures that can be far more valuable. Watch for beach development work. When pipelines are being laid and seawalls, breakwaters and piers constructed, work these areas of excavation.

'Reading' a Beach

You must develop the skill needed to "read" a site. If you learn which features are important and which are not, much of your battle is already won. As you research records, histories and old maps, be on the alert for clues to landmarks and locations. Reading a site requires recognition of key features and the forces that may have acted upon them over the years. Because beaches protected from winds that cause large waves are more popular than unprotected beaches, more lost treasures can be found on them.

Popular beaches usually feature fine, clean sand with a wide and gradual slope into the water. Remember that changes continually occur as a result of both man and nature. Popular play areas of yesterday may scarcely be recognizable as beaches today. Find such a section of "lost" beaches and hunt it profitably. Not all are still connected to the mainland; some are now separated by lagoons and marshland. Some have been converted into bird and wildlife sanctuaries.

As seashore development has increased, former swimming beaches have disappeared or been permitted to erode. New business and industry have been permitted to take precedent over recreation and natural beauty. Breakwaters, harbor extensions, jetties and damming or otherwise diverting streams and rivers have destroyed once-popular play areas. Treasure lost there years ago, however, will remain forever — or, until it is found. Search out these treasure vaults and reap a harvest.

As experience accumulates, you will discover "mis-located" treasure in areas away from people. How did this happen? Perhaps this is where people used to congregate; it might once have been a swimming beach. Then, for some reason, the old beach was abandoned along with its treasure. Another reason is natural erosion caused by tides, surf and such that redeposits objects. Even though such action is seldom permanent, always keep in mind the forces that cause it to happen — and, watch for them in action. These forces do not occur accidentally, and they can create treasure vaults for you to find and empty.

Obvious other places to search for beach treasure are man-made spots. Walk onto a beach and observe people at play. Watch children of all ages as they frolic. Then, when they tire of that activity, watch them scoot away. Coins fall from pockets...rings slip off of fingers...bracelets, necklaces and chains fall into the sand as young people play their games. Other more subtle games are being played on beach chairs and blankets, but wherever people relax, coins and jewelry are certain to fall into the sand.

Search around trails, walkways and boardwalks. Never pass up an opportunity to scan the base of seawalls and stone fences. People without chairs often camp by these structures where they can lean back. Always search under picnic tables and benches. Sure, you'll find lots of bottlecaps and pulltabs, but you will also find coins, toys and other valuable objects. Search around food stands, bath houses, shower stalls, dressing sheds and water fountains and under piers and stairs. Posts and other such obstacles are good "traps" where treasure can be found.

Equipment & Clothing

A piece of equipment often overlooked by the beginning beach hunter is the digging tool. Why, it would seem that almost any type of digger can be used in loose beach sands...even hands! I strongly counsel against this for several reasons, the first and foremost of these being the abundance of broken glass. In fact, I recommend gloves, at least one for the hand that gets into the sand at all. Another reason for not depending upon hands alone as a digging tool is that the beach hunter cannot always expect to find targets in soft beach sand. Sometimes a sharp and hard-edged digger will be absolutely necessary.

In dry and loose sand, however, sandscoops are reasonably good. A quick scoop, a few shakes and there's the find. In wet sand, however, such tools are just a waste of time. It takes too long to work damp sand out of a scoop, except in the water where onrushing surf can flush wet sand.

Other gear needed for beach hunting includes an assortment of pouches, a secure pocket for storing especially good finds and a place for personal items. If you've hunted for treasure at all, you probably already have some ideas about recovery pouches. Let me offer just a couple of suggestions for the beach:

– Place *all* detected items in a pouch; don't stand in the surf trying to decide whether an object is worth keeping. Carefully inspect your finds occasionally and discard trash properly. When I find an especially valuable article, I return to my vehicle or camp to stow it properly.

– Use care in handling rings with stones. Often, mountings corrode during exposure. Examine jewelry with your pocket magnifier; when a mounting shows corrosion, handle that ring with extra caution.

– A fastener on a pouch is not a necessity on the beach unless you lay your pouch down carelessly or let it bounce around in your car.

– Pouches should be waterproof to prevent soiling your clothes and sturdy enough to hold plenty of weight.

– Many pouch styles can be mounted on a belt. I often wear a web-type military belt carrying a canteen and an extra pouch or two.

Concerning clothing, the best advise is to dress comfortably. But, protect yourself against the elements you're sure to encounter on the beach. Obviously, you'll want to keep warm in the winter and cool in the summer, but I caution you to shade exposed skin areas to protect against sun and wind burn. In warm weather I wear shorts or lightweight trousers, a light (but long-sleeved) shirt, socks and comfortable shoes or sneakers. I wear a wide-brimmed cotton hat with some sort of neck shield. Even when hunting only on the beach, I'm always prepared to get wet. Sometimes an attractive low place in the sand will be yielding recoveries, and I must be prepared to follow it right into the water. More about clothing and equipment in Chapter 13.

Family beach outings become even more enjoyable (and profitable) when a metal detector is taken along to the seashore.

Winds, Tides and Weather

Wouldn't it be great if the ocean suddenly receded several feet, leaving your favorite hunting beach high and dry? You could walk right out and recover lost treasure so much more easily. Well, the ocean does recede slightly every day during low tide. This exposes more beach to be searched and also makes more shallow surf area available.

Listen regularly to weather reports and forecasts to learn of prevailing winds. Look for tide tables in your newspapers. Strong offshore (outgoing) winds will lower the water level and reduce size and force of breakers. Such offshore winds also spread out sand at the water's edge, reducing the amount that lies over the blanket of treasure. On the other hand, incoming wind and waves tend to pile sand up, causing it to increase in depth. Pay attention to winds and tides, especially during storms.

Storms often transfer treasure from deep water vaults to shallower locations. Plan a beach search immediately following a squall. If you are hardy enough, try working during the storm itself. Always remember that extremes in weather, wind and tides can cause unproductive beaches to

Almost any quality detector will perform adequately on the beach, but special instruments are required for surf and deeper waters.

become productive. Storms play havoc with beach sands. Fast-running currents that drain a beach can wash deep gullies in the sand...gullies to bring you closer to the blanket of treasure.

Look for tidal pools and long, water-filled depressions on the beach. Any areas holding water should be investigated since these low spots put you closer to the blanket of treasure. As the tide recedes, watch for streams draining into the ocean. These locate low areas where you can get your a detector's searchcoil closer to the treasure.

Scanning Tips

Do not race across the sand with your searchcoil waving in front of you. *Slow down!* Work methodically in a preplanned pattern. Unless you are in a hurry and seek only shallow, recently lost treasure, reduce scan speed to about one foot per second. Let the searchcoil just skim the sands and keep it level throughout the length of a sweep. Overlap each sweep by advancing your searchcoil about one-half its diameter. Always scan in a straight line. This improves your ability to maintain correct and uniform searchcoil height, helps eliminate the "upswing" at the end of each sweep and improves your ability to overlap in a uniform manner, thus minimizing skips.

Don't ignore either very loud or very faint detector signals. Always determine the source. If a loud signal seems to come from a can or other large object, remove it and scan the spot again. When you hear a very faint signal, scoop out some sand to get your searchcoil closer to the target and scan again. If the signal has disappeared, scan the sand you scooped out. You may have detected a very small target. It might be only a BB, but at least you'll know what caused the signal.

Remember. Your quality metal detector will *never lie to you*. When it gives a signal, something made of metal is bound to be there.

The matter of trash on a beach is one that daily becomes more urgent to all of us beachcombers. I refer especially to plastic trash that is more than just unsightly. Fish and sea birds become entangled in six-pack rings; sea turtles mistake plastic bags for jelly fish and swallow them; birds peck at plastic pellets and try to feed them to their young. Similar harm results from countless other plastic items that are carelessly discarded on our nation's beaches every day. What can a beachcomber do about it?

Most hobbyists carry out the metal trash they dig because we all benefit from its removal. But, what about non-metallic trash? Certainly, none of us carries around a container large enough to hold all the plastic trash and broken glass we find in only a few hours. Let's join together to help, however, and dispose properly of as much trash as we can. We perform a service not only for all beachcombers and sun worshipers but for sea creatures and bird life as well. How about it...can't we join together and help one another?

Sea Stories

One of the great thrills of beach hunting is that "big one" that always awaits...your chance to strike it rich. I'm very serious when I suggest that you should always be on the alert for treasure stories and legends. Please don't waste good money on a "treasure map," but don't ignore those tales of missing treasures...of great losses and "almost" or partial recoveries. Not only will this add excitement to your hobby, but the stories sometimes prove to be true!

Before you spend too much time seeking the mythical "pot o' gold," however, you should attempt to verify the sea story you are following. Major concerns before you get yourself seriously involved in tracking legends are, first, to make certain that the treasure ever actually existed and to what extent it was recovered. Then, you must locate the precise spot where it is rumored to have been lost or where it was only partially recovered. Remember that beaches run for miles and that names of various areas can change regularly. Also, the appearance of beaches change. Erosion may take years to alter a beach radically, but storms can transform its appearance in just a few hours.

Some Final Tips

Schedule your beachcombing expeditions according to current (hourly) weather reports. Stay alert to weather forecasts and go prepared to withstand the worst.

Plan your treasure hunting expedition. Make a list of all you will need the day *before* you make the trip and check all gear carefully before you leave.

Always put *batteries* at the head of your list (see above). And, always check your batteries first if your detector should stop working. Some hobbyists take these longer-life batteries for granted and expect them to last forever. Believe me, they won't. You'd be amazed at how many *broken* detectors can be "repaired" with new batteries.

Take along a friend, if possible. If you go alone, leave word where you'll be. Always carry identification that includes one or more telephone numbers or persons to call (with coins for the pay phone taped to the list). Your personal doctor's name should be on this list.

Be wary of driving in loose sand. Carry along a tow rope and a shovel. You may need someone to pull you out of trouble, or you may have to dig ramps for your wheels if a tow vehicle isn't handy.

If there are no regulations to the contrary, you may want to search among crowds. But, don't annoy anyone. Angering the wrong person can result in immediate trouble, or you may find a complaint filed against you personally and the hobby in general. You certainly wouldn't want to cause a beach to be put off limits for metal detecting.

Whenever possible, return any find to its owner. Try to oblige when someone asks your help in recovering a lost article. It might be feasible for you to loan them your detector and teach them how to use it. Who knows? You might add a new member to our brotherhood. When helping look for a lost article, it's a good idea to keep its owner close by throughout the search so that they will know whether you succeed or not. If you can't find the article, get their name or address; you might find it another day.

Do not enter posted or "No trespassing" beaches without obtaining permission. Even in states where you are certain that all beaches are open to the public, do not search fenced or posted areas without permission. Never argue with a "loaded shotgun;" leave such property owners to themselves.

Finally, remember that a modern metal detector is a wonderful scientific instrument. It searches beneath the sand, where you cannot see. It is always vigilant about the presence of metal. But, no detector can "do it all." You must develop powers of observation that keep you attentive to what a detector cannot see. Watch for the unusual! Sometimes you'll visually locate money, marketable sea shells or other valuables. The real benefit of developing keen powers of observation, however, is to enable you to enjoy the glories of the beach to their fullest and never to overlook the signposts pointing to detectable treasure.

There's treasure to be found near the water! And, vast amounts are waiting...enough for all. I sincerely hope that you'll join the rest of us beachcombers in searching for this lost and hidden wealth. When you do, perhaps I'll see you on the beach!

There's still another vast area of this wonderful hobby we haven't yet discussed. It includes the forgotten treasures that have been left behind in ghost towns and those that can be found on old battlefield. I'm especially fond of this kind of hunting and I want to tell you about it in the next chapter.

Chateau in the Alps...
A Hobbyist's Dream!

alk about a detector hobbyist's dream! How about a chateau with 29 rooms in the French Alps...a chateau whose construction was begun in the 15th century? That's what Arnold Anderson found when he took his GTA 1000 along on a visit to his daughter's home in Europe.

"What a place," Arnold related. "During World War II the chateau was owned by a Russian who used it as a hide-out for refugees fleeing the Nazis."

And what a place for his GTA 1000. On the chateau's grounds he found more than 100 coins from nine different countries with dates that spread over four centuries. He dug up projectiles from World War II and found many thimbles lost by women who had been sewing under the trees.

Arnold's oldest coin was dated 1610, and he found Swiss coins from the 18th century and several Swiss, French and Italian coins from the 19th century, including a dix centime piece featuring a likeness of Napoleon III.

Discrimination capabilities of the GTA 1000 made finding the coins easier since Arnold could block out everything with conductivity lower than a nickel, yet still have a cursor show him every target...a real help in searching for coins of so many different metals!

Forgotten Treasures

*W*hat do you see when you encounter an old, deserted house? I always visualize a potential treasure trove because abandoned old houses are rarely *empty*. And, often, the items they contain will prove to be valuable to someone who can find them with a metal detector.

Searching individual structures or ghost towns encompasses all phases of the hobby. Looking for caches, seeking coins, hunting for relics...all of these are part of searching buildings, cabins or other structures that were once occupied. And, you'll be surprised at how many old buildings and cabins are just waiting for you and your detector.

Any place where people once lived or conducted business will produce treasure that can be located by a metal detector. Thousands of abandoned homesteads, stores and commercial establishments, schools and churches as well as townsites, forts and military installations await you. The list of places where people "used-to-be-but-no-longer-are" is truly endless.

And, many of these locations have never been searched! We have discussed more than once the importance of not being intimidated by the fact that a specific location has been searched before. Remember that metal detector capabilities have improved dramatically in just the past few years and that the proficiency of individuals can vary widely...even with the best and most modern instruments.

When searching such a location, you may one day be hunting in an area that contains only a few relics. Then, the next day, you might encounter an entire town; that is, structures still intact with buildings and rooms in them just as they were when people — for some reason — left, perhaps decades ago.

To search abandoned properties properly you must learn the techniques of hunting outdoors. But, you must also learn how to hunt in structures of all kinds. Never forget that incredible treasure caches have been located in the walls, floors and ceilings of old buildings.

When you're searching a building with a metal detector, also keep in mind that most wooden structures contain a truly countless number of nails. You can expect your detector to respond with multiple target signals. Of course, you don't want to tear into a wall just to locate a nail. We recommend, therefore, that you use only enough discrimination to reject troublesome small nails. You'll not be likely to overlook a large money cache!

Simply hearing a signal will often let you know in which range your target falls. Always inspect the graphic target display to see if it agrees with the audio, which will usually be the case. There will be times, however, when only a faint blip might cause you to suspect a target should be rejected, yet your display indicates it to be made of metal with high conductivity. Dig to determine if the target is junk such as a large stove lid (which might be a relic itself!) or something that has simply overridden the electromagnetic reporting capabilities of the searchcoil.

When scanning around window and door frames, be alert for signals you receive from the iron sashes used to suspend the window frames. Don't rip open a wall looking for treasure until you have exhausted all techniques for peering into that wall by other means. Most wall areas can be visually inspected simply by pulling slightly back on a single board and shining a flashlight into the cavity. Never tear down or otherwise destroy anything.

In fact, you should leave all structures in better condition than you found them...without harm or defacement of any kind. Walk away from buildings and cabins that you have searched, leaving them in such condition that no one can really tell whether you found treasure there or not. Destroy nothing. Do not tear out any boards that you cannot replace easily. Walk away leaving all sites in such condition that no one can really tell whether you found treasure there or not. Use common courtesy at all times. Remember, you might want to return! Another rule to remember is always to ask for permission before searching any property. Remember that every structure or square foot of ground in the United States is "owned" by somebody.

Most old houses and buildings will present detector hobbyists with a seemingly endless amount of junk iron of all shapes and sizes. If you are looking for coins, brass objects or other similar targets, therefore, use a small amount of discrimination. It would take weeks or months — perhaps, even years — to dig every metal target to be found around an old farm house. This makes discrimination mandatory. But, let me urge that you employ all techniques we have already considered...the smaller Super Sniper searchcoil, minimum discrimination, slow scanning, careful study of questionable targets, etc.

Because of all the junk you are certain to find you must develop techniques for properly identifying targets before you dig. If you're using a detector that has a visual target display (and, I hope you are), rely heavily on it to do this for you. And, try to learn more about how your detector identifies targets both audibly and visually. Try to correlate the audible and visual signals before making a decision on digging a specific target.

In areas with lots of junk targets your audio may sound often with blips and other sharp signals as well as an occasional coin tone. If you've followed my advice and are hunting with one of the new computerized instruments, remember that your detector will be identifying deeper targets than others. Consequently, it will give you more signals over a given spot of ground than another detector that is not searching as deeply. You have several options that will not only cut down on spurious sounds but identify all targets more accurately. First, you can reduce detection depth (sensitivity) somewhat and still probably have sufficient depth for most targets. If your detector has an adjustable threshold, you can reduce it almost to silent to reduce bogus signals. Try various combinations of adjustments such as these to achieve optimum audio for any difficult ground you encounter.

What about searching houses where people still live? Provided they are old enough and have enough "history," occupied homes can present many targets of which current occupants are unaware. You can look for caches hidden and forgotten by previous dwellers. You can seek jewelry or silverware that was hidden for safekeeping and never recovered for some reason. You can find coins and other items of value that might have fallen through cracks to rest under floorboards or between walls.

Truly, treasure waits to be found wherever men and women have been. Humans misplace, lose and hide items of value in locations where they can be discovered only by a modern metal detector. Of course, never fail to use your eyesight in searching any location. What can be discovered simply lying on the ground or on the floor of a structure is amazing.

When you are searching an old cabin or house or hunting anywhere else in a ghost town, however, you should understand that surface items were probably picked up by relic and antique hunters long, long ago. You will need a quality metal detector because most the type of objects you seek are beneath the ground or are concealed from sight in some other way.

Relic Hunting

Popularity of searching battlefields, ghost towns and other areas for lost relics has been growing in recent years among metal detector hobbyists. Perhaps this is due to the greatly increased abilities of the new detectors. Perhaps it's because of the surprisingly large prices now being paid for relics. Perhaps it's just the natural growth of the hobby. Nevertheless, when you become successful at finding coins, you may want to expand your interest to relic hunting...especially if you live in an area where such objects abound.

You'll find that relic hunting is not considerably different from cache hunting which is covered in the following chapter. Perhaps one difference is that caches are usually larger than the targets relic hunters seek, such as a single button or spent projectile. Techniques of scanning and locating, however, remain the same.

In the literal sense of the word "battlefield relics" are scarce indeed in the United States. Since the Civil War of 1861-65, our great nation has been blessedly free of wars fought on its soil. Even before then, the only actual war "battles" fought in what is now the United States were those in the East Coast states during the American Revolution and the War of 1812 and a skirmish or two in Texas during the Mexican War.

But, "war battlefields" notwithstanding, there are relics aplenty to be found. The richest trove of all, of course, is located in the Southern states where so many Civil War battles and skirmishes occurred. Encounters with Indians also left vast quantities of relics to be found throughout what was once frontier country. In addition to these obvious "battles" were many other actions of arms that resulted in battlefield relics that can be found by today's metal detectors.

And, the term "relics" includes far more than implements of war and destruction. In fact, the dictionary defines relics as "a trace of the past." Collections specialize in relics from almost every sector of society and life. For example, when you're searching "out West," don't neglect antique barbed wire that is found by your detector. Just a single strand of old fencing can give you a real history lesson when you research it. And, it can be valuable to a collector as well.

What size searchcoil should you use when hunting for relics? You may be thinking that if you seek a small bullet, you shouldn't use the large 12 1/2-inch searchcoil. Not necessarily! Large searchcoils are often used by veteran relic hunters. Of course, extra skill is required. So, sharpen your skills with larger coils, and you'll be rewarded with deep buttons, coins and projectiles that have been overlooked by others. Because you will need all the depth possible when searching battlefields, especially when the battles were fought years ago, we urge you to use the larger searchcoil.

Because so many battle sites are in low-lying or swampy areas, it is well to make certain that your searchcoil is submersible. Don't be confused by such designations as "splashproof" or "waterproof." You will want a searchcoil that can be completely submerged a foot or so...to the cable connector that attaches to the control housing. If you doubt your searchcoil's capabilities, ask the manufacturer. All Garrett searchcoils, of course, are guaranteed submersible.

You perceptive hobbyists already know what comes next, but I'll repeat, anyway: For maximum depth and sensitivity, use headphones and set your audio controls for the faintest threshold you can hear. This advice has been proven worthwhile over the years.

As worthwhile as the following advice...scan slowly. Your goal will be to cover an area methodically, completely and thoroughly. So, slow down!

Most experienced relic hunters use no discrimination. Here's one reason why: If a valuable coin is lying right next to an iron object and you are using discrimination, the iron object may cancel out the coin, causing you to miss it. Too, there are valuable iron relics that you'll want to find.

Because any battlefield might contain explosives, take all necessary precautions. Any time you dig into an object and you suspect it might be an explosive, consult an authority…quickly! Remember that many loaded guns are lost and that ammunition that has been in the ground a number of years can sometimes still be fired. Don't get into arguments with explosives of any kind. It might prove dangerous to your health!

A military historian recently told me that he owed his life to a Garrett detector. While scanning in Korea, he often found remains of Chinese soldiers still bearing live hand grenades. The advance warning provided by his Garrett permitted him to call in a bomb squad to dispose of the explosives.

Research

How can you locate areas to search for relics? Your answer, again, is research…research and more research. Often, all the research in the world can't answer the question of exactly what battle action occurred in a particular area or whether gunfire took place at that precise location. Only your detector can prove the locations of battle or gunfire by locating cartridges, bullets or other spent projectiles.

Over

Civil War relics such as these projectiles are the quest of many hobbyists, and there's no better way to find them than with a quality metal detector.

Facing

Sightseeing and treasure hunting are combined with exploration of a ghost town and the search for valuables and relics left behind a century ago.

Research is particularly important since so many of the "obvious" locations to search for relics are now located in various state and national monuments and parks where the use of metal detectors is either banned entirely or highly restricted. You'll need to seek out locations about which only you and few others know...perhaps you can develop these by learning more about the history of your ancestors.

Why Search for Relics?

Unlike the cache hunter, who is searching primarily for monetary reward, there are other reasons to search for relics and battlefield souvenirs. Some relic hunters are always looking for evidence to prove history, while others seek significant objects to add to a personal collection. Of course, many relics are sold, some for surprisingly large sums, and most relic hunters search for all of the above reasons.

It is fascinating to read tales of the early day settlers of this nation who simply picked up relics while plowing crops or discovered them under brush in old battlefields. Except for an occasional find, those days have passed long ago.

Over

Memories of the Old West are brought to mind by this pair of antique handcuffs found with a metal detector in a mountain ghost town.

Facing

With the Depth Multiplier two-box searchcoil, a metal detector can be used to discover large and deeply buried relics.

A Family Pastime

Searching old buildings and cabins in a ghost town is a pastime that you can combine with a family vacation. I know that the Garretts do this often as we travel throughout the U.S. Perhaps your travels take you to unusual and interesting places...old abandoned towns and mining areas. Can you believe that thousands of tourists travel through such areas without stopping? Yes, even metal detector hobbyists who don't realize all that they are missing! Relics, old coins and other valuable items are just waiting to be found. In many of the areas that especially cater to tourists, you will find libraries and visitor facilities especially equipped to suggest places for you to search with a metal detector. These free sources of information are generally more accurate than hand-drawn maps and guides you receive from individuals.

Don't forget all the many things that you have learned about research, however! Use these techniques to the fullest. Remember how helpful old-timers can be and what you can learn from long-time residents. Even when you seem to be receiving little information from local citizens or are treated almost rudely, try not to get discouraged. Don't give up! Good people exist, and there is always the library. Keep looking for "that spot" where you can find treasure with your metal detector. Persevere, and you will be successful.

Do your research homework to locate lost and forgotten ghost towns. Find them, and then search thoroughly. It will pay financial dividends as well as enable you to relive history and have fun doing it.

In the next chapter I'll discuss a specific type of forgotten treasure...buried money caches.

They Didn't Trust Banks in the Old Days

*I*n a quarter-century of visiting southwest Arkansas Claude Odinot was intrigued with the tales of hard Depression times...and the stories about people who distrusted banks.

"They talked of one person in particular who always seemed to have plenty of money on hand back in the 1930's but never had a bank account. Yet after he died, his family found little cash. Of course, speculation centered on buried money caches."

None had ever been found.

Claude finally joined the hunt with the purchase of his first detector, a Grand Master Hunter CX II. He searched long and hard; his detector found a lot of metal, but little treasure.

Then Claude spotted a knoll profiled behind the house, standing between it and a small draw. "What a place for a money cache," he thought, and his CX II showed its agreement by giving a loud and clear signal. Four inches down he found a stove lid and broken glass. Below it was a jar that might have been inside a larger jar. It appeared to be filled with dirty water, but something else was inside...150 old coins, mostly silver from the New Orleans mint.

Claude is now confident of finding even larger caches nearby...and he's searching diligently.

Money Caches

*M*ention treasure hunting to most folks, and they'll soon visualize cache hunting...whether they know it or not! They'll probably think of digging up a treasure chest on a desert island, finding loot hidden in a cave by outlaws long ago, recovering money stashed away by a hermit who didn't trust banks, discovering valuable relics. It's the thought of locating *big* treasures that entices most people to become interested in cache hunting. But, in most cases it's the *little* caches that are usually found...perhaps ten times as often as the large ones.

Whether you want to get into this aspect of metal detecting is something you must decide for yourself. I hope you'll give it a try for reasons other than the monetary rewards. It's in cache hunting that you must use your brain a little more to research a project. And, because carrying out your search often taxes you in many ways, the rewards seem to be more worthwhile.

Cache hunting is *different*. Always remember that...no matter how successful you become at finding coins...no matter how much jewelry you dig out of the beach. Looking for money caches generally means searching for a larger quantity of buried treasure. Your cache can be an iron kettle filled with gold or silver coins. It can be a cache of gold or silver bars or even guns. You will generally be looking for objects much larger than single coins...though smaller money caches in tobacco cans are sometimes found. Cache hunting is about finding big prizes...and, usually finding them pretty well hidden!

Will your detector have to be different for success in cache hunting? Not necessarily. Most quality detectors can do a good job detecting small caches to a depth of two feet or so. Of course, if you're going for something really deep, you should have an instrument with true Deepseeking all metal capabilities proven to be capable of detecting deeply buried caches. And, imaging lets you know target size.

You might also want to learn something about the two-box searchcoil called the Depth Multiplier which enables you to find large, deep targets. This attachment multiplies by several times the depth-detecting capability of any compatible detector . An important feature of the Depth Multiplier is that it will not detect small objects, thus enabling it to overlook little pieces of junk metal.

This chapter, though, is designed to explain how you can successfully use a quality detector to hunt for a cache. First of all, you must think and act *differently* when you hunt for a cache. Always remember you'll be looking for *big* (relatively speaking) *money.* True, you'll need all the knowledge you've developed in other kinds of hunting. And, your basic techniques may be the same as those you've learned to use. It's your overall manner of searching for a cache — from research to recovery — that must be different...if you are to be successful.

Unless you're searching for a cache in a building —
where you know that it cannot possibly be too far away —
always use the largest searchcoil possible. Remember that
larger searchcoils can detect larger objects and detect them
at greater depths. Money caches have been found at all
depths (within an arm's length seems to be most frequent),
but you want to be prepared for extremes. In some outdoor
areas, where washing has occurred and drainage patterns
have redesigned the landscape, caches have been recovered
from depths different from when they were originally
buried.

What is a Cache?

Caches come in all sizes, and they're generally dreamed
of as a Well Fargo money box, a big trunk or a set of saddle
bags...all stuffed with gold coins, old bills, antique jewelry
and the like. I sincerely hope that this describes the cache
you locate some day. In the meantime, please remember that
most caches are small. They consist of a tobacco tin holding
a few bills or a quart fruit jar filled with coins...maybe old
and worth more than face value, maybe not. Such little
caches aren't as exciting, perhaps, as the Wells Fargo box
or those outlaw saddle bags, but valuable nonetheless.

Regardless of the size cache you seek, you must not take
a chance. So, use a large searchcoil. There is no doubt that
even the best treasure hunters have left behind deep caches
that were beyond the range of the finest detectors available
in earlier years. These caches await you and other hunters
with the 21st-century instruments capable of finding them.

When searching a farm yard for a money cache, look
closely at specific objects and obstacles in that yard, such as
a well, the corners of the farmhouse and its chimney. Search
inside the chimney and all outbuildings...especially those

that housed the dogs, chickens and other animals that make loud noises when disturbed.

Never fail to search an old garden area. Here's where the farmer's wife may have hidden some "rainy day" savings in a fruit jar. Remember that when people buried caches, they didn't want to be observed. It would be quite normal for a farm wife to hide a jar of money in her apron, carry it to some special location in the garden and "plant" it secretly.

Bury A Cache

Successful searching for caches requires considerable experience...and thinking. You must learn to put yourself right in the shoes of the person who hid that cache for which you are searching. It's easy to understand why a person wouldn't just run out into his yard haphazardly and dig a hole to bury a jar full of gold coins. No, if you were burying a cache, you'd select a secret place and a secret time to bury it...perhaps, at night during a storm. And, your "secret place" would be one that *you* could find in a hurry while others would overlook it!

Practice this yourself. Put some money (or something similarly valuable) in a jar. Go outside your house and bury it. That's right. *Go ahead and bury it...* if only for a few minutes. After you've done this, you'll be able to ask yourself the questions that probably occurred to that person who hid any cache you ever seek.

Would you do it in broad daylight? Would you just walk out into the yard and start digging? Probably not, because you wouldn't want anyone to see what you were doing. So, choose the right time and the right place to bury your cache. Can you find it easily? Can it be found accidentally by a stranger? Will it be safe? Many other questions will come into your mind as you recover your own cache and relocate it a time or two. This is good experience that will make you a better cache hunter.

When you hear a story or get a treasure map about a cache that is buried high atop a mountain or in some other difficult-to-reach location, you'll ask yourself such questions as, "Why there?" Why, indeed would someone have climbed a high mountain or scaled a steep ravine to bury a cache?

Try to learn the thinking of someone who is burying a cache, and you'll have better luck finding it. It won't be just "luck," either! Whenever you're tempted to attribute the success of another cache hunter to "luck," remember what the old football coach said when they accused his team of being lucky: "We had to be there for the luck to happen!"

When searching for caches always try to use detectors that employ the latest technology. It is truly amazing how much more effective today's modern instruments are than those with which we were so well satisfied just a few years back.

Research

Most cache hunters spend a major portion of their time in research, seldom mentioning their work except to another cache hunter. Since proper research can require extensive travel, expenses necessary simply to determine the *existence* of a single cache can be considerable even before a detector is assembled and turned on. Sometimes, cache hunters are required to pay sizable sums to obtain information. Often, groups share the cache on a percentage basis, a common practice also for gaining permission to search on private property. Occasionally, a special detector must be purchased because of the nature of the ground where a cache is sought. Proper financing, as well as patience, is often required.

Successful cache hunters are a dedicated breed, but their single-mindedness pays off in tangible rewards. They are willing to overcome obstacles because they are seeking real treasure — financial wealth — and a bundle of it! In fact, dedicated cache hunters perversely welcome the obstacles since they limit the number of hobbyists in the field searching for the same prizes.

Of course, even the good ones aren't successful every time. The beginner should realize this and not become discouraged. We advise working on several projects simultaneously. Since research and travel can be expensive, it is good to "double-up" on the uses you can make of it. Always remember that there are literally millions of dollars stashed in the ground waiting to be found. If you persist, sooner or later you will hit a cache. It may be only a few hundred dollars tucked in a tobacco tin; then again, some treasure hunters have become wealthy from pursuing this fascinating occupation.

Cache Hunting Basics

Because cache hunting is different, the basic concepts governing it are also somewhat different than those of other forms of treasure hunting. Consult the basic treasure hunting rules on Page 45, but also consider these guidelines that have proven successful for most of us cache hunters:

– Conduct *extensive* research; you can never know too much about your target and the individual(s) who hid it.

– Be *patient* throughout your effort, from planning to scanning to recovering…even after you dig up your prize.

– Never *assume* that because your target may be big that it will be *easy* to find. Sure, some cache targets are quite large. But, they may be deep as well and, thus, more difficult to locate.

– Do not *anticipate* anything about the cache or where it is located. You can be certain that "things" will not be the way you imagined them.

By all means, remember to use all those special tricks that have proved so successful for you and your detector. As I continue to emphasize in all of my books and articles, basic techniques of metal detecting remain the same because the laws of physics do not change. It's the manner in which you apply the basic techniques that determines whether you can be successful in cache hunting.

Let's consider some factors that will govern your application of all basic techniques of treasure hunting. Each of these factors can enter into successful recovery of deeply buried caches:

~ Geographic location of the treasure site; you *must* locate it.

~ Ground condition of the site and vegetation covering it; if you have a weed-cutter, use it.

~ Mineral content of the soil; a computerized detector will remove this frustration.

~ Physical size of the cache (generally overestimated!); remember that any size searchcoil can find a large and shallow cache.

~ Depth of the cache; it's best to use the largest searchcoil available for your detector or a deepseeking two-box coil.

~ Changes that might have occurred at the site *since* the cache was buried (generally not considered!); fill dirt may have been piled over it; a trash dump may now be there. Natural changes may have been even more dramatic; layers of silt may cover it; erosion may have changed the entire appearance.

~ Your detector and its searchcoil; don't try to go hunting elephants with a BB gun.

Misjudgment of any one of the above can keep you from recovering the prize(s) you seek. Experienced cache hunters always make allowances for the condition of the search area and the fact that their cache may be both deeper and smaller than anticipated. Pay close attention to the description of where it was buried. And, when you reach the probable location of your cache, don't rule the site out just because of its present-day appearance. So what if it *doesn't* look like that description written decades or centuries ago! Remember that trees and shrubs grow taller or can die and be removed entirely. Plus, you should never underestimate the effects of both erosion and sedimentation. What was once a deep ditch might be just a depression today...and, vice versa.

Take your time. Be patient, and reap the rewards.

As you may already know from coin hunting experiences, the longer a coin has been buried, the easier it is to detect. Depending upon coil characteristics and other factors, freshly buried metallic objects can be detected to about one-half the depth of the same objects when buried for a longer time. This same phenomenon holds true in the detection of buried caches.

Unless you are confident that the treasure has been found, never pass a site because you have been told that it has been worked before. You don't know who searched or when or with what kind of detector! Also, I'm convinced that no matter how often a site may have been searched over the years, more treasures were missed than were recovered. In researching a Louisiana cache I returned to a location where I found only a deep and empty hole instead of the treasure I expected to dig. As I reinspected this hole, it occurred to me that maybe the *real* treasure had been buried even deeper, with only a sampling of items left in a container above to satisfy anyone who might accidentally stumble upon this site!

Concerning "worked-over" sites, just consider the old parks where coins continue to be found year after year after year…and, not all of them newly minted coins either. These parks never seem to be completely hunted out. Newly designed detectors seem to be performing miracles. Now, consider the rugged, highly mineralized terrain where many caches are found and consider also the eternal question of just how deeply they were actually buried. These caches are far harder to find than coins. Remember, also, that anyone who searched a site in past years probably did so with a detector whose capabilities are far exceeded by your modern computerized instrument.

Never forget what a tremendous advantage today's detectors give you over the so-called "old pros." I sincerely believe that even a relatively inexperienced treasure hunter with a new computerized detector can search an area more effectively than the most experienced veteran cache hunter who insists on using obsolete BFO or TR detectors popular in bygone years.

No matter how much skill an old timer had, he could not possess the *scientific abilities* of our modern metal detectors.

Recovery

Since most hobbyists don't get involved with a cache that requires a bulldozer or backhoe, a long-handled shovel is the primary recovery tool. I also recommend a long steel probe that you can use to save time…where soil conditions permit. If you believe that your detector's response indicates a target large and deep enough to conform to the cache for which you are searching, you can probe the spot before digging.

Most experienced cache hunters go to great lengths to avoid calling attention to themselves. One way to do this is to carry detectors and all other equipment into the field in a backpack. You then appear to be just another hiker. A large backpack will usually accommodate a 12-inch searchcoil, along with small shovels, your detector's housing and the other tools necessary for an average recovery.

There are numerous reasons for not calling attention to yourself or your search for caches. First of all, you're looking for money; 'nuff said. Plus, you'll be busy and won't need the attention of even honest curiosity-seekers. And, if word ever gets out about your recovery of a cache, you'll be amazed at the number of people who will try to take it away from you...legally by claiming rights to all or a part of it...or, simply, by force.

On private property or if there is a question of ownership, negotiate an agreement with the property-owner or individual(s) who might have a legitimate claim to your cache *before* you begin searching. And, never put your trust in a *verbal* agreement. A wise man once said that verbal agreements aren't worth the paper they're written on. Also, never leave an open hole after you have discovered something. Even a landowner with whom you have an agreement can get excited about a large hole. He visualizes it filled with gold coins, and trouble may lie ahead.

When you are working with partners, make certain that all arrangements are made in writing *before* you start spending money on research and equipment and, certainly, before any cache is discovered. Many of us have had unpleasant experiences, particularly in working with inexperienced treasure hunters, such as landowners.

Again, my advice to a cache hunter is to keep a low profile in every way. Don't call attention to yourself. Pay your legitimate taxes. Insist on your rights...in a quiet, yet firm, manner.

If you haven't experienced the enthusiasm of cache hunting, you can't know what you've missed. True, you'll relish the same exciting joy of discovery and receive the benefits of relaxation, fresh air and outdoor exercise that other forms of the hobby provide. But, you'll give yourself the marvelous chance of making that *really big* discovery...the awesome thrill that should come to everyone at least once in a lifetime.

In fact, this subject of cache hunting is so interesting and exciting, I plan to write an entire book on it!

In the next chapter I'll explain why the hobby of treasure hunting with a metal detector is truly not only one for the ages but one for *all* ages!

Gold coins were the prize when this money cache consisting of a coin purse was discovered by a detector in the rock foundation of an old house.

5-Year Old Is Already Detecting Veteran

*W*hen he was only five years old, Paul Sabisch was already a veteran treasure hunter. He even gave lessons in using his GTA detector to sandbox pals. But, this achievement wasn't surprising, for while literally in the cradle, Paul was accompanying his parents, Andy and Roseanne Sabisch of Canton, Georgia, on treasure hunting excursions. So, after he acquired the necessary walking and talking skills he naturally demanded "a 'tector of my own."

Andy could well understand since he had begun detecting himself at the age of six and today is nationally recognized for his treasure hunting prowess and also ranks among the leading metal detector journalists. Her discovery of four gold coins says plenty about Roseanne's detecting prowess. So, the parents

This small pocket detector is being used to search an abandoned house, "looking" behind the walls where a money cache is suspected to be hidden.

purchased an easy-to-use OneTouch detector, cut a few inches off the upper stem and Son Paul was in business.

How did the lad do? His father explains, "In a park with a large play area Paul refused any help and said he knew what he was doing. Well, he did! I watched as he began scanning. Almost immediately he got a solid signal and, after scratching around in the woodchips, proudly held up a shiny new quarter."

Twenty minutes later Paul emptied his pockets of $2 in change, an earring and a matchbox truck. Paul's not selfish. That first earring and all jewelry he's subsequently found have been given to his mother. But not the coins! With those he buys for himself some of those toys that his parents might refuse him.

Children & Seniors

*T*his delightful hobby of treasure hunting with a metal detector knows literally no bounds. Children of all ages and both sexes enjoy finding treasures, of course, but boys and girls can also learn to appreciate the subtleties of the hobby at a surprisingly young age. Automatic computerized detectors are so easy to use that even a pre-schooler can turn one on and find treasure with little or no supervision.

As for so-called senior citizens, many of them tell me that the hobby of metal detecting is improving the quality of their lives in retirement. Hunting with a detector provides exercise and fresh air while adding an air of excitement to an existence that could become routine.

Children

After you achieve just a little bit of proficiency in finding coins and other metal treasures with a detector, you will learn that your new hobby can enable you to entertain your own children or grandchildren or the neighborhood boys and girls with whom you are friendly. In fact, when they discover that you know how to find coins with a metal detector, young people will flock to you. Whether you like children or not, you're sure to become more attractive to them.

Children seem to have a thirst for adventure and a real curiosity about how things work. They can be fun to have around and always have energy to spare. When you hunt in public parks or on public beaches, you'll soon find yourself becoming a veritable Pied Piper with a band of youngsters following your every move. Why, to them you're an honest-to-goodness magician with a magic wand that finds money! You can use this attraction to bring happiness by sharing the hobby. Little people will enjoy hunting with you, and you can take pleasure in their joy as you introduce them to this wonderful hobby.

You'll be surprised at how quickly youngsters learn how to use your detector...especially if you're hunting with a OneTouch instrument.

Let them turn it on; then, show them how to scan properly. Of course, they'll be in a hurry to run across the park with it. Aren't young people always in a rush? So, show them how to scan methodically and help them to inspect every signal carefully; first by studying the graphic target indicator; then, by digging properly.

I hope you'll avoid the temptation to let your little friends become only your "diggers." They'll probably love doing it — for a while. But, it's really not fair. They deserve some of the fun of using the detector, and they may not come back for another day of only digging!

Unfortunately, the attraction detectors seem to have for children can occasionally prove somewhat of a hindrance. You'll encounter adults who wonder just why little boys and girls are following you around. And, sure enough, some children are going to ask for coins or other objects that you find. Now, I don't like to be a Scrooge — especially about pennies — and you probably don't either. But, it's been my experience that giving away objects while you're still hunting will only attract more children who might really become nuisances. If you want to surrender coins or other finds, wait until after you've finished detecting.

Even then, you must be careful just how you give away the coins you've found and who you give them to! Remember that offering anything, especially money, to strange children can often cause trouble. The gesture can so easily be misinterpreted. Today's rules of conduct require that you be especially careful in associating with young people who are strangers to you.

So, expect children to flock around that magic wand you're scanning across the park or over beach sands, and be sensible in dealing with them. Yet, remember that you can let boys and girls enjoy the fun with you and really intensify it for children, grandchildren or other young people who are your friends. It's just another aspect that makes the hobby of metal detecting so attractive.

Seniors

Variety, excitement, expectation, adventure...within limits — each of these is proposed as a prescription to older individuals for living a more satisfying and longer life. Now, the methods for following these directions, depending on physical capabilities, budget requirements and individual interests, can be varied — to say the least! That's why you'll find so many books and articles giving advice on how to grow older successfully.

Each man or woman of any age needs a good reason for meeting the day with vigor and expectation. This is especially true of older individuals because goals lend direction and vitality to the person pursuing them. A day with no goal, not even the goal of creative loafing, is condemned to be a day of aging and nothing else. A life without goals is, by definition, no life at all.

During our early and middle years most worthwhile goals are thrust upon us by mentors, business conditions and similar situations. We accept good goals without even thinking about them. As we grow older it is less likely that worthwhile goals will be "thrust upon us." We must take fate into our own hands.

I hope you will agree that the hobby of metal detecting satisfies much of the advice usually offered to older individuals that is designed to make their lives more meaningful and satisfying. Now, I'll admit that I'm prejudiced. Yet, let's consider some of the suggestions for happiness ordinarily given to older people:

Put variety into your life is a common recommendation, followed by numerous suggestions — some of them rather outlandish — for achieving this variety. These proposals make me shake my head and chuckle. What greater variety is there to be found than by listening to a metal detector and digging up all that it signals!

It's an accepted fact that unless you are stimulated...unless you are challenged, your mind will stagnate. I'm not talking about artificial simulation with any of the fancy so-called "therapies." I recommend something practical and obvious like metal detecting to set your mind to working again.

Create excitement is another common bit of advice given to seniors, usually accompanied by some rather bizarre suggestions. Once again, I merely smile and think of the excitement I experience every time my metal detector finds a target!

Look forward to something, oldsters are told…have a sense of the future. Many studies have shown the value of keeping a fix on the future as a means of maintaining mental health. It figures, of course, that people who look forward must have a sound sense of themselves and their place in the world. When my detector sounds off on a target, I certainly know where I am. And, you can bet I'm looking forward…to digging up what my detector has found. Wouldn't you? I'm also looking forward whenever I research a long-lost cache. I can't wait for the day I am able to seek it with a detector.

Summing all of this up is generally this piece of advice that is given to older men and women: *find adventure.* Easier said than done for most! Yet, treasure hunting with a metal detector *provides all the adventure that I need.* It offers armchair adventure which comes from reading about hidden treasures and researching their locations. Then, I can decide how much more adventure I want (or need) by looking for one of them. Believe me, adventure will seek out anyone who hunts actively with a metal detector.

I've heard it said that the goal of every man or woman should be to die young…at the oldest age possible! Does a "fountain of youth" exist? Of course it does — it is inside each of us! Only, not everyone knows how to find it. I propose to older men and women that they permit the hobby of metal detecting to help discover their own fountain of youth.

As you've been reading this book, I hope you've thought about where you expect to find treasure. In the next two chapters we'll discuss systematic methods for helping you locate such places.

Old Maps Lead Hobbyist to Treasure

*M*y metal detector buddies told me not to waste time here. They said the area had been hunted out. But I hadn't hunted it with my CX III, and I wanted to hear what TreasureTalk had to say!"

What James Brewer didn't tell his buddies was that because he had already found a 1935 Lincoln cent here at eight inches, he knew he had a good site. You see, Jim realizes that a metal detector hobbyist must hunt in places where people might have lost things. And, his waybill to treasure is a 1938 map of Livingston County, Illinois, that gives him the location of schools, most of which were demolished decades ago.

His present location is an empty field today, but a school was located here back into the 19th century. Scanning near the old foundation he found interesting items such as square glass ink wells, green copper conduit pipes and lightning arresters.

Then, he hit the jackpot! His CX III led him to a fruit jar filled with Indian Head cents, including one from 1866, silver half dollars and quarters and a 1921 Morgan dollar.

Research

*W*here do successful hobbyists *find* all those coins and rings and things? That's easy! They find them anywhere people have been—which is practically everywhere. And, they learn just *where* (and, when) people have been through research. Most newcomers to the hobby think that research is dull work; they'd rather "get into action" by scanning (often haphazardly) with a detector. And, they may be wasting both time and opportunity by looking in the wrong places.

Many people tell me,"Oh, I have so little time for the hobby; I don't want to waste any on research." Actually, the right kind of research helps you *save* time and can greatly increase the value of your take. Although it's true that research can be work, it pays off! Research is profitable in many different ways, and to me it can be as much fun as the actual treasure hunt itself.

To repeat a truth, successful treasure hunting can often be 99% research and 1% recovery. Don't think of research as though it were an uninteresting stint in the back room of some dusty, dimly lit library where you must pore over volumes of scarcely legible books, articles and newspapers. Research can be fun. It can become something you enjoy and look forward to. In fact, research for many hobbyists is more than just adding to your store of knowledge about the past — enjoyable as that can be. It means simply keeping your eyes and ears open.

Potential hunting sites are still virtually limitless. Finding a good one now simply requires more imagination and more research that it once did...which merely makes the hobby more interesting. And, it should also be more profitable! Good research can lead all of us to better sites, and computerized OneTouch instruments will recover those old and deeply buried treasures that might have been left behind by hobbyists with older or inferior detectors.

After you get into this hobby (become obsessed, some might say), you'll continually think about it. You'll scan newspapers and magazines for stories and data about local sites. When you talk with people, especially oldtimers, you'll ask them about such-and-such a place. You'll ask them if they remember whether the present park and grandstands are located where they were decades ago. You'll ask them if they remember incidents when individuals lost jewelry or other valuables.

There's an unlimited amount of research information available to you. The only limits will be those you impose upon yourself. Knowing that everyone has shortcomings, you should never rely entirely upon the work of others. When someone is willing to write about or talk of a treasure, you can be sure that person has abandoned the search for it for one reason or another. You must analyze all treasure data with a cautious eye. Failure of another person to complete the research and recovery could have been due to lack of funds, time, or, simply, interest. But, if a hobbyist traveled to a site and investigated it, that person must have believed in the story.

Treasure found by accident represents but a small percentage of that located by persons using good, acceptable research practices. Without proper research you'll be as lost as a driver without a map in a strange city. You need a waybill...directions to guide you to the best locations. These waybills, these directions, come from many sources, both public and private. And, the "waybill" may be simply something you have observed!

I *never* advise either buying a treasure map or taking the word of a single individual as gospel. If possible, you must always find the primary source or at least check your facts with more than one person. To begin at the beginning involves a study of basic research material and sources. You must know *what* you are looking for and that it exists. Certain forms of treasure hunting require a knowledge about where specific types of treasure, can be found. You don't search for pieces of eight in a city park or seek lost gold rings in a child's sand box. (Even though tiny children's rings and mother's rings are occasionally found there!)

There's no one-two-three-step research procedure. One hundred hobbyists may look for one hundred different treasures one hundred different ways! The main thing is to *get started* by defining these goals, which you can call...

A Treasure Check List

~ What are you looking for?

~ Can you prove it exists?

~ Where is it?

~ Will you have clear title to it if you find it?

~ Have others looked for it?

~ How do you know they didn't find it?

~ What will it cost you to find it?

~ Is it worth what it will cost you to find it?

Certainly, these questions are rudimentary, but yet very important. They may appear to apply just to "really big" prizes, but the same questions pertain to coins in the park or an earring lost in your back yard. What I want to emphasize is that you not go searching for some mythical pot of gold. Spend your time wisely and efficiently. Don't waste even a minute looking for treasure unless you are sure, based upon your research, it exists.

Once a person has begun the fascinating hobby of hunting with a metal detector he will no longer need convincing that treasure is waiting to be found. He will soon have the problem that other hobbyists face—simply too many places to search.

You'll probably start out searching for coins, and I urge you to start at home. The first place every hobbyist should begin searching is his own backyard. The mistake too many individuals make when they set out to hunt is ignoring nearby treasures while they hasten off in search of Captain Kidd's buried chest or that fabled Wells Fargo box hidden by Butch and Sundance.

I can't tell you how many people have complained to me that there is nothing in their area worth searching for. And, they're talking about their entire neighborhood or town! The truth is that it would take an army of thousands of hobbyists, working many years, to search and clean out all the productive areas existing today in any city or county. And, by that time, think how much additional treasure would have been lost and would be waiting to be located!

You'll soon discover that so-called "hot spots" can be found in just about any given area. In other words, people have congregated in some places more than others. It's true today; it's always been so. You can learn about "hot spots" simply by observing people as they go about their daily routines. Drive over to your school or college campus. When you are at church, watch the people coming and going...see where they stand and talk, where their children run and play, where cars let out passengers. Just a little common sense and observation will enable you to increase your finds.

Your imagination will be called on when you investigate areas that are no longer used by the public. From your own experiences try to visualize where the crowds would have gathered. Then, develop "criss-cross" metal detection techniques that will let you sample the particular areas you have selected. As you search different areas, keep accurate count of whatever you find and where you found it. From this data you will soon learn the probable location of the best places to search. Some hobbyists seem to be able just to walk into an old park and immediately begin scanning their detector over the exact spot where coins are to be found.

Use your ears! Make it a habit to listen to the old-timers. Talk to family, friends, storekeepers—especially the older, retired people of the community who were postmen, bus drivers, merchants, policemen, firemen and the like. These individuals will have a wealth of information that can help you locate valuable hunting areas. And, don't ask them simply where you can find old coins. Just try to start them talking about the past...their activities and pleasures of yesterday.

What about nearby parks? How old are they? Are they in the same place now they were 50, 75 or 100 years ago? Check the official records of your city. Investigate vacant lots. What used to be there? As you drive through rural communities and small towns, stop to talk with the old-time residents. And, listen to them! You'll be amazed at how many are just waiting for an audience. Let them tell you where the general stores, saloons, banks and cafes used to be. Let them tell you of all the things that people used to do. That old-timer won't know it, but he'll be telling you where lost treasures are waiting for your metal detector.

One excellent method for finding the oldest objects is to locate geographically where your town or city was located when it was founded. Often, the present center of the town's activity is far removed from the original center of population. For example, the original town of San Diego, CA, is a few miles north of the present downtown area. When my family and I visited "Old Town" San Diego and talked with George Mroczkowski, he showed us where he had uncovered many artifacts and valuable coins.

Florida treasure hunters determined the location of the original site of Tampa, FL. Many relics and coins, including Spanish reales, half dimes, large cents and other denominations were found there. The American coins are dated in the 1830s. Even older coins could still be found because the original town of Tampa was erected just outside Fort Brooke's walls in 1823.

There is no doubt that most hobbyists truly enjoy getting out into the parks, playgrounds and other outdoor areas. Searching for lost coins and valuables simply adds to the pleasure. And, they delight at every discovery! But, let's face it—the finds that we enjoy *most* are those with the greatest value. Selling them lets us benefit financially from

This senior proves with today's prizes that treasure hunting with a metal detector is a hobby that knows no age limits.

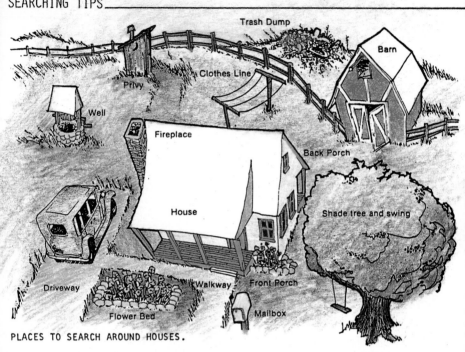

PLACES TO SEARCH AROUND HOUSES.

the hobby, and they give us more pleasure if we display them in a collection.

Granted, each of us experiences a thrill when we dig up anything valuable, even a coin worth just face value. But, I hope that each of you shares my special joy in finding anything old...a remnant of the past. The half cents, large cents, two and three-cent pieces, half dimes, Liberty-seated quarters...and especially *gold coins*. These are often worth many times, often thousands of times, their face value...yet they provide an additional historical and romantic thrill. Such coins are seldom located in parks, around school yards or in areas of relatively recent occupancy. You must look for older habitations and areas of activity to find these older and rarer coins. This is where *research pays off*!

Make it a daily habit to read the lost and found sections of newspapers, no matter where you find yourself. Quite often, lost valuables will be advertised with a reward for their return. Make prior agreements concerning your remuneration or reward and then locate these lost valuables. Newspapers, too, are filled with information of the locations of public congregations (company picnics, family reunions and other such gatherings).

Spend time at the local library or newspaper office reviewing the issues of yesteryear. You will discover the

Above: At this farm site there are easily 100 potential places where treasure could be hiding. How many can you name?

Below: Scanning this Maryland farm occupied since the 1700s produced such treasures as the old sleigh bell, matching those on a harness still in use.

location of old parks and playgrounds, band concert sites,fairground and circus lots, along with information on public activities that occurred in the past. Notices of lost articles are also to be found here. Only your enthusiasm will limit your efforts and your success. And...if you are a dedicated treasure hunter, you will discover many heart-fluttering stories that suggest veritable bonanzas. May you find some of them!

Bulletin boards in laundromats or other public areas often contain notices of lost items. Check these places frequently. and consider posting your own "Have Detector...Will Travel!" notices on these bulletin boards?

It's a good idea to make contact with local police and insurance agents or even insurers and law enforcement officials outside your area. Tell them you have a metal detector and that you are willing to help them locate lost jewelry and other valuables. You may be surprised at the services you can perform, as well as the rewards you will receive. This is just another example of how the owner of a metal detector can increase his annual income with only a reasonable amount of thought and physical effort.

There is no way that I can stress strongly enough that importance of the ideas presented in this and the following chapter. Sure, treasure is just waiting to found nearly everywhere, and I am certain that each of you can find a goodly amount. Still, the maximum monetary value in treasure hunting and the greatest personal rewards come from finding the old, valuable coins...and, finding them as a result of your own research and hard work. You can only grow even more enthusiastic as your rewards increase!

Treasure is *what* and *where* you find it! And, I'll get even *more* specific about the where in the next chapter!

Potato Field Yields Treasure Trove

*S*he inspected the Nottingham potato fields daily. It was plowing time, and Jean Ward of Mansfield, England, wanted to be the first hobbyist this season to search these fields which had already yielded so many Roman coins.

When she located a field that was ready, she found the farmer eager to have her Garrett detector inspect it since his son had lost a part from their potato machine there while plowing. Furthermore, he reported the promising news that this field had been the site of an ancient battle.

Jean studied aerial photographs before going to work. Her first finds were good...Victorian coins, George II, George III, even a hammered silver coin of Henry III, numerous musket and pistol balls, perhaps from that battle...but no Roman coins.

Jean persevered by studying the field more carefully and selecting a particular location. This hard work was rewarded when her Garrett found a treasure trove of Roman coins that had probably been stored in a bag or receptacle. She found 29 denarii, most from the 1st and 2d centuries. Her prize was purchased for display in a local museum.

And, guess what? She also found the part from the potato machine!

Where & When

*H*ere are some general ideas about where I and other hobbyists have found treasure with a metal detector. This material is adapted from a chapter that was in the First Edition of my *Successful Coin Hunting* more than 20 years ago. And, the material is as pertinent (and, popular — I still get compliments on it) today as it was then. The list is by no means all-inclusive, nor will it *ever be complete.* By the time you've read a page or two you'll recollect and add a favorite spot or two.

Where People Live(d)

Indoors

Closets and shelves.

In the walls.

Above and beneath door and window sills.

Underneath or along baseboards.

Underneath or along edges of linoleum or other flooring...especially adjacent to holes.

Garages and storage sheds.

Outbuildings, such as barns and animal shelters.

Crawl spaces under structures.

Outdoors

Your own yard.

Driveways and parking areas...where people would have gotten out of cars or carriages.

Doorways...where coins might have been spilled.

Next to porches and steps...where people might have sat.

Porch and step railings...where children might have played.

Around and along all walkways and paths.

Around old outbuildings...on the ground where they stand (or stood) and along the path to the house.

Around hitching posts and hitching post racks.

Between gate posts.

Near mail boxes. Remember that in rural areas people used to put coins in the mailbox, along with the letters to be picked up by carriers. Often, some of these coins would spill when pulled out by the postman. I've heard tales of coin hunters finding Indian head pennies and other old coins around the locations of rural mail boxes that haven't been used for years.

Well and pump sites.

Storm cellar and basement entrances.

Around watering troughs.

Along fence rows and around stiles.

Under large trees...children could have played or had swings here..."shade tree mechanics" might have worked here...especially seek trees with adjacent built-in benches.

Under clotheslines or their possible locations.

Around patios and garden furniture areas...look here, also, for permanently installed benches and seating areas.

Along trails, especially those leading to outbuildings.

Where People Play(ed)

Fishing piers, boat ramps and landings.

Ferryboat loading and unloading sites.

Fishing camps and health resorts ("watering" places).

Abandoned resort areas.

Horse and hiking trails...especially spots where people may have stopped to rest or camp.

Swimming pools or "holes"...especially abandoned or old ones.

Children's camps...especially concession or play areas.

Around ski tow loading and unloading areas.

Beach swimming areas.

Miniature golf courses...driving range tees.

Shooting and target ranges (but, expect to find lots of shell casings).

Old springs or wishing wells.

Pioneer campgrounds.

River fords.

Bluffs and embankments that might have served as playground "slides."

Abandoned trailer parks.

Beneath stadium seats.

Bandstands, gazebos, entertainment platforms...or where they once stood.

Amusement parks, fairgrounds, carnival and circus sites...the potential here is unbelievable.

Rodeo grounds...today's and those of the past.

Old horse-racing tracks and spectator areas.

Parks/recreation areas...and, here you can let your imagination run wild...benches, drinking fountains, large trees, steps, picnic tables, sports areas, walkways...this list could go on and on.

Drive-in theater locations...around concession areas or where children might have played...around ticket windows.

Motels...current and abandoned locations...recreation areas and concession machines.

Historical markers and highway locations displaying maps or otherwise presenting good photographic possibilities where travelers might have stopped.

Tourist Spots

Just thinking about historical markers and the roadside parks where tourists and travelers stop should cause you to recollect numerous other tourist-type places where coins are sure to await your metal detector:

Tourist stops of **any** kind...wishing wells and bridges, hilltop lookouts, scenic spots.

Below all of the above or anywhere people might have pitched coins for luck.

Near litter cans on highways (but expect cans and pulltops).

Footpaths and resting spots along trails or roadside parks.

And, Did You Think About?

Anywhere people congregate and anywhere people have been.

Around service stations, particularly older or abandoned ones.

Around old churches...where people might have gathered to visit after services...where they took "dinner on the grounds"...where children played.

Revival meeting sites.

Schools and colleges...around playground equipment, bicycle racks...in front of water fountains and doors, where students waited in lines.

Don't forget old or abandoned schools. In the East Texas area where my wife Eleanor was reared is the old Steele Academy, a training school for boys that closed its doors before the turn of the century. This is good coin-hunting territory.

Where were the old schools and academies located in your area a century ago...150 years ago? How about the CCC camps and military training grounds of 50-60 or so years ago? Talk to old family members, friends and old-timers. Jog their memories about where people *used* to congregate.

Old highway cafes and truck stops...drive-in eating was a popular pastime throughout the United States for many years. Lots of coins were lost at such places.

Ghost towns...along the boardwalks...in the streets.

Around ladders and fire escapes permanently attached to buildings. As a boy, I recall playing on a slide-type fire escape from a church. I'm sure that I and others lost coins here.

Anywhere cars were parked...at sporting events... revivals...market areas.

Old stagecoach stops, relay points, trading posts.

Bus stops...school bus stops.

Around telephone booths.

Outdoor taverns...look for those with loose gravel.

Around flea markets and auctions.

The secluded "lovers' lane" spots.

Courthouses and other public buildings...around benches on the lawn, paths and walkways.

Are you beginning to understand that there is really no limit to the many types of locations where treasure awaits your metal detector? I urge after that you study my list, you try to recall your own personal experiences over the years and keep your mind open to the many additional locations where you might find lost treasures. My list is really just a beginning that should suggest a lot of additional locations to you. I'll bet you can come up with many that are not listed above.

Listen and Learn

There are unlimited numbers of people with whom you can talk to learn where valuables might have been lost or are being lost *right now*. Consider the types of persons on the following list and, once again, add to it from your own background:

Old-timers of all kinds head the list. And, you may be on it yourself! Yet, let your friends and acquaintances talk about the past and the way things were "then."

Caretakers of parks and recreation centers.

Lifeguards at swimming areas. They can perhaps tell you about jewelry whose loss was reported.

Police and personal property insurance agents can also serve as sources of tips on lost valuables.

Highway clean-up crews know where people normally congregate...that's where they will lose things; they may know about losses that have been reported.

Clergymen, especially older ones, can give you the locations where outdoor revival meetings were held in "the good old days."

Bus drivers (look for retirees) sometimes know of long forgotten communities they once served.

Construction crews...especially those tearing down houses or buildings. Always follow the bulldozers! As soon as they shut down for the night, try to be there with your detector...but, *obey* all safety rules and don't trespass.

Historians, amateurs preferably, who can relate local history and point out where people once used to gather, transact business and play. They can sometimes provide the location of once-populated but now deserted and forgotten community sites. A common occurrence is for the town sites to "wander;" that is, to move away from where the community was founded. Valuables lost long ago won't "wander;" they are waiting for you to dig them up!

"Listen" to history books written about the areas you are interested in exploring. They are unbeatable sources of local information. Read all the books you can on local history. You'll learn from each of them. No matter what you *think* you know about an area, seek all the information you can obtain from others.

Above all, don't assume that books have never been written about your area. Check with libraries and historical societies for manuscripts. Don't forget the chamber of commerce. With each history book—amateur or professional—that you read, you will arm yourself with additional knowledge of dozens of places that can possibly yield large quantities of old and valuable treasures.

When to Hunt

Newcomers to the hobby often wonder about just *when* to hunt. Is there a special time better than others? Time of day? Time of year?

The answer to all these questions is to hunt any time...day or night, morning or evening, rain or shine, summer or winter...all seasons are treasure hunting seasons. Use your own best judgment and, always remember, you're hunting because you *want to* and because you *enjoy it*. If the hobby should ever become tedious or boring, give it a rest and wait for your interest to return. Frankly, I can't imagine that hunting for treasure could ever be boring, what with all the wonderful coins, items of jewelry and other valuable objects just waiting to be found.

But, when to hunt...here are some suggestions. During lunch breaks, scan nearby parks. After your evening meal, you might go to a nearby park or swimming area to search for an hour or so. On weekends you can spend as much of the days as you like searching outlying and out-of-town sites. On vacations make it a habit to stop along the roadway at various parks and roadside stops where you and your family can stretch your legs and refresh yourselves. At the same time, you can search and recover a coin or two lost by those folks who have used the park before you.

Treasure hunting is more than a good excuse to arise in the morning. It lets us limber up and get the blood circulating. Get up an hour earlier than normal; drive to the park or into any public area and search along deserted traffic paths that are heavily congested during the day. Get out before most people do; the rewards will be yours. Any time you're driving along and see an area that looks promising...stop, get out with your detector and scan a sweep or two. You'll never know what you may find until your detector sings out and you dig. Many hobbyists tell me that they *never* go for a walk now without taking a detector along!

Now, use this chapter to stir up your own imagination. You'll quickly discover that there's truly no limit to the places where you can use your detector to find treasure!

And, in the next chapter I'll try to answer questions you might have about clothing and equipment for this hobby.

European Hobbyists
Know What 'Old' Means

*W*hen a metal detector hobbyist in the United States digs a coin or artifact from the 19th century, the find is a real antiquity. Even an item from pre-World War II days is considered "old."

Yet Geraldo Aranda of Zaragoza, Spain, with his GTA 1000 regularly finds relics and coins that date from before the time of Christ. Such is the lot of the metal detector hobbyist in Europe!

Modern metal detectors are finding relics that archaeologists have only dreamed about, for these new computerized marvels have "eyes" that can see far beneath the soil.

No, Geraldo will assure you, it's not as simple as pressing a touchpad, making a few scans and digging. Dogged research, hard work and patience are required for the good finds...such as the payroll of a Roman army, 20,000 bronze coins, found by a Garrett detector or centuries-old Greek and Roman helmets that have been discovered.

And, more and more American hobbyists are joining in the fun by taking their Garretts with them on European vacations!

Clothing & Equipment

*S*urprisingly little, actually! What must I wear? Am I dressed properly? What will *they* think of my outfit? These questions should never trouble any metal detector hobbyist. Concerning equipment, you must have a quality detector; that's an absolute must. I also recommend headphones. And, you'll want to have some sort of tool(s) to dig up your discoveries, plus a pouch or container of some sort to stow them. Beyond that...whatever pleases *you* is all that you need.

One of the most persistent and amusing questions I receive concerns "what I wear" when I'm treasure hunting...what brand of shoes! "I want to hunt just like you," hobbyists often tell me. As if clothing could make any difference! My big advantage is the high-quality Garrett detectors I use and the abilities that I have developed to get the most from them.

As far as what I wear...why, I wear just what I told you two paragraphs back — whatever pleases me. There are no standards to follow, or expectations any of us have to fulfill. Every hobbyist is on his own. And, isn't this wonderful?

Clothing

Comfort is absolutely essential to the enjoyment of the hobby. Thus, comfort and protection (plus modesty, I guess) should be your major clothing considerations. Clothing must always offer protection while permitting considerable freedom of movement. During warm seasons, you can wear just about anything that pleases you. Yet you may have to protect yourself from environmental conditions other than stormy weather. I'm talking about such things as the hot sun, cactus or thorny bushes.

Certainly, head and skin protection are the key summertime considerations. I wear a scarf and sometimes a neck shield that attaches to my hat. Often, I wear soft, cotton gloves. Sometimes I take a poncho along in case of rain. A poncho is good, except it often gets in the way, especially when you stoop to dig. During really wet seasons, a two-piece lightweight rain suit is better; but it can be hotter, especially if the top half cannot be unbuttoned.

More often than not, I enjoy a little rain, whether I'm treasure hunting or jogging. During my four years in the U. S. Navy aboard a ship that was continually traveling from one climate to another, I learned that weather is a friend, not an enemy. I cringe when I hear TV weather reporters say, "Well, folks, we've got some pretty drab, nasty and terrible

This treasure hunter has protected himself well against the elements, and he will also be careful to protect the control housing of his detector.

weather in store for you." Frankly, *all* types of weather are a wonder and a joy to me. In fact, getting out into the weather is one of the pleasures of the metal detector hobby.

Since you'll be on your knees digging, you might try ready-made or home-built knee pads. Thick pads or home-made varieties that you cut out of sturdy material prevent skin abrasion and sore knees. Unfortunately, there are two disadvantages to wearing knee pads. If the bands become too tight, they may restrict blood circulation or otherwise cause discomfort. The second disadvantage is that dirt and sand may work its way inside the pads, soiling your trousers. Commercial knee pads are made of tough neoprene with a velcro elastic strap. You might try making your own by cutting two lengths of rubber from an automobile innertube. The sections, each couple of feet in length, are worn over the pants' legs centered at the knees and can be held in place by rope or strapping that comes up and attaches to the belt or clothing.

When working in the sun, which is usually quite often because the shade trees seem always to be *over there*, I rarely wear anything more than light trousers, a T-shirt (or, a light, long-sleeved shirt when the sun is really strong), appropriate protective footwear, a wide-brimmed hat, neck covering and soft cotton gloves. I use lots of sunscreen, number 15 or higher. In areas where you may encounter cactus or shrubs

The sturdy pouch on the right hip of this hobbyist enables him to store coins and other finds safely without having to worry about soiling his apparel.

that can scratch, I urge you to protect your arms and legs. Today's "little scratch" can become tomorrow's infection. Sometimes I wear polarized glasses that reduce surface reflection of the sun.

I wear different styles of World War II military headgear which I purchase at Army/Navy stores. Other personal articles include a canteen (I use the military type on a web belt), snacks, digging tools and toilet paper.

As far as footwear is concerned, it depends entirely on the topography. Whether it's boots for the desert or mountains where snakes may be encountered or aqua-shoes on the beach, protection and comfort are the two criteria. Remember that if you plan a serious session with your detector, you'll be on your feet for a long time.

Cold weather must be considered. When the thermometer drops, out come long pants and long-sleeved shirts. Insulated boots are good, especially if you have cold feet. A seaman's type wool cap and gloves can be life savers.

Underclothing, or thermals and thick socks will keep you warm. Jogging suits and insulated underwear such as that worn by cross-country skiers offer another option. Select insulated articles only after some experimenting. If your inner clothing soon becomes saturated with perspiration, you've dressed too warmly. Several thin layers that let moisture escape while trapping air are the best combination.

Remember, when clothing becomes wet, it loses most of its insulating properties. As noted earlier, water-protective gear must be ventilated to let moisture escape. You may have to do some experimenting, but give cold weather hunting a try. As long as the ground isn't frozen, you'll like it!

Headphones

Other than your digging tools there's only one accessory that I urge you to use regularly. That's a set of good headphones. You should *always* use headphones whenever you search with a detector. They are especially useful in noisy areas, such as the beach and near traffic (when you are looking for dropped coins around parking meters). Headphones enhance audio perception by bringing the sound directly into your ears while masking "outside" noise interference.

There is no question that most people can hear weaker sounds and detect deeper targets when quality headphones are used. As proof, bury a coin at a depth that produces a faint speaker signal. Then, use headphones and scan over the spot. You'll be amazed at how much better you can hear the detector signals with headphones than you can with the speaker alone.

Most detectors are operated at full volume with the tuning (audio) control adjusted so that a faint sound (threshold) continually comes from the speaker. When a target is detected, the sound volume increases quickly from threshold to maximum loudness. Headphones allow this threshold to be set even lower, giving improved performance.

Headphones also can aid by contributing to your privacy. They help you keep others from knowing what you are doing because a person standing only a few feet away will not be able to hear the headphone signals unless the sound is very loud. Spectators will have a hard time asking you questions, too, with your ears covered by headphones.

Carrying bags and suitcases are optional pieces of equipment that a person may wish to buy. There are many different types and styles on the market. Some are made of flexible vinyl plastic and others are rigid like a suitcase. The latter provides maximum protection for your detector regardless of whether it is stored or transported.

A *coin apron* is certainly a must in any kind of coin hunting or for any other kind of light treasure-seeking activity. These are generally constructed of waterproof material with at least two pockets, one for good finds, the other for junk. All junk items, of course, should be collected for later discarding. Never leave trash lying about! When I don't use an apron, the coin and treasure pouches I carry are very sturdy and usually have a snap or zipper/velcro fastener.

Batteries, of course, are essential to your success with the hobby. Always carry a spare set with you whenever you hunt…no matter how fresh you believe the batteries in your detector to be. And, if your detector ever "fails," check its batteries first. You'd be amazed at how many instruments are returned to the Garrett factory for "repair" when all that's needed are new batteries.

Called NiCads (from their nickel/cadmium composition) *rechargeable batteries* are now reasonably reliable and long lasting. You can save money by using them, especially if you hunt with your detector often. Follow manufacturer's recommendations on keeping the batteries in a recharged condition. Special NiCad Rechargeable Kits are offered for many detectors, including the Garrett GTA and CX models.

In the next chapter we'll talk about metal detectors — the instruments themselves — and choosing the *right* one for you!

He and GTA 'Smell' The Good Stuff'

*C*alifornian Vic Benamati will never see the age of 80 again, but he'll also never see a site that doesn't offer some appeal to him as a metal detector hobbyist. Vic claims that between them he and the Garrett GTA on which he depends can "smell the good stuff, especially quarters." His treasure hunting pals don't argue with him. This veteran hobbyist with his careful selection of treasure hunting locales produces results that speak for themselves.

Living in an area of California settled between 100 and 150 years ago, Vic and his GTA have discovered many coins, artifacts and pieces of jewelry which he modestly describes as "pretty nice."

A recent feat in which he takes particular pride was the discovery of three valuable coins in a residential yard. How valuable were they? The quarter was dated 1857 and two half dollars were 1862 and 1871, all minted in San Francisco.

"Oh, how I love the sound of money...that thrilling music that comes from my GTA when it detects a coin!"

Metal Detectors

*S*electing the *right* metal detector will be one of the most important steps you take in pursuit of the hobby of treasure hunting. And, unless you get your hands on that *right* detector, you aren't going to know all the enjoyment that is possible with this hobby.

Significant improvements in the performance of detectors themselves ... just in the past few years ... have spurred the recent growth in popularity of the hobby. And, I believe that one of the most significant improvements has been the successful effort to *simplify* the operation of detectors. That's right...simplify. As performance of the instruments improved, their operation became more effortless.

Today's metal detectors are almost unbelievably easy to use. Touch a single control on a modern computerized instrument and begin finding treasure immediately. Why, you may *never* have to touch another control. It's that easy to hunt with today's OneTouch instruments!

Yes, that's all there is to it! Moreover, visual target indicators on some detectors report on every target your searchcoil sweeps over. That's right, you *see* every target you encounter, and you get an audible alert on those worth digging. *Imaging by Garrett* even discloses a target's size!

Now, I'm talking about quality instruments...not the inexpensive "toys" that have discouraged so many people because they just wouldn't find much of anything. And, that's what we've been discussing throughout this book...quality metal detectors that are, in truth, scientific instruments. They're not inexpensive, but the subject of cost is dealt with later in this chapter. Let me just say that I've been using my own high quality detectors to find all sorts of things for some 35 years, and I'm still amazed at the ease with which our newer models can be operated. Automatic OneTouch detectors are so simple that it's sometimes uncanny.

Modern computerized circuitry will find *more* treasure and will find it quicker and easier than even the best of yesterday's instruments. Over the years Garrett has participated (led the way, some tell me) in the continuing progress that has been made in detector circuitry and design. Year by year we steadily improved all of our instruments. But, the advancements made by the Garrett Engineering Lab in just the past few years have been truly breathtaking. You see, metal detectors have entered the computer age, and treasure hunting has never been simpler than with a computerized OneTouch detector with imaging.

How They Work

It's not necessary to understand the scientific principles of metal detection to hunt with a modern computerized detector. Touch one control, and you can find coins, rings, jewelry, gold nuggets, caches or whatever you are searching for without understanding how your detector is operating.

It's always helpful to remember, however, that a good metal detector will never lie; it will simply report what it detects beneath its searchcoil. Thus, successful hobbyists find it useful to know just how a metal detector goes about finding metal and telling them what it has found.

I like to explain what a detector *is* by pointing out what it is *not*. It's not an instrument (Geiger counter) that detects energy emissions from radioactive materials. It's not an instrument (magnetometer) that measures the intensity of magnetic fields. It doesn't "point" to coins, jewelry or any other kind of metal; it doesn't measure the abundance of metal. A metal detector simply detects the *presence* of metal and reports this fact.

And, it detects metal essentially by the transmission and reception of radio wave signals. When a detector is turned on, a radio signal is transmitted from its searchcoil (antenna), generating an electromagnetic field that flows out into any surrounding medium, whether it be earth, rock, water, wood, coral, air or some other material. The extent of this field, and the distance (depth) to which a coin or other metallic item can be detected depends upon the type of searchcoil, power used to transmit the radio signal and the resistance of the medium into which the signal is transmitted. Simply stated then, metal detection occurs when the electromagnetic field encounters metal.

This creates electrical eddy currents that flow on the metal's surface, and generation of these currents causes a power loss in the electromagnetic field. Another effect occurs when the metal's presence distorts the detector's electromagnetic field and changes its shape.

Circuitry of the metal detector is designed to sense this loss of power and the resulting distortion of the electromagnetic field. The detector simultaneously interprets all these sensations and signals to the operator that some type of metallic object is present. Numerous other factors, such as ground mineralization and atmospheric conditions, can also inhibit or enhance detection. And, some metals are better conductors than others, thus generating more currents to be detected. If you're interested in learning more, I recommend my book *Modern Metal Detectors*. As my editor Hal Dawson delights in pointing out to me, this book may tell you more than you want to know about metal detectors!

Discrimination is a popular word with detector hobbyists that can be defined simply as the elimination of unwanted targets. Modern OneTouch detectors have simple basic discrimination preset at the factory to keep you from finding most bottlecaps and similar metallic trash. Of course, it is possible to regulate this discrimination to your precise requirements.

Depth of Detection

How deep can a metal detector find metal? When an electromagnetic field flows out of the searchcoil, several factors determine whether detection is possible: electromagnetic field strength, target size, surface area of the target and the type of metal in the target. How far the electromagnetic field flows from the searchcoil also depends on the size of the searchcoil, quality of its construction and materials that are present in the earth. Larger searchcoils produce a more extensive field that can penetrate more deeply to detect deeper treasures.

Of the factors that determine how deeply a target can be detected only the electromagnetic field and the circuitry to interpret its disturbances are a function of the detector itself.

Two other important factors, size and surface area, are determined by the individual targets.

Simply stated, the larger a metal target...the better and more deeply it can be detected. It is realistic to expect that a coin-sized target can be detected under normal conditions to depths of six to nine inches. Yet, detection is often inches deeper and, sometimes, shallower...all because of such variables as ground mineralization, moisture in the soil and the conductivity of the metal in the target itself — or even how it is lying in the ground.

How Detectors "Report"

When a treasure hunter is scanning his searchcoil over the ground or in the water, a detector reports information on targets in various ways:

– Increases in audible volume (universal on all detectors);

– Graphic information presented visibly;

– Meter deflections (types of meters can vary greatly, along with the amount and accuracy of the information they present).

Good targets will generally be announced by a clear and distinct tone while the detector will produce "blip-like" sounds for pulltabs and similar small trash.

Learn to listen closely to your detector's signals, and develop the skills to interpret what it is "telling" you through its LCD and sound indicators. I have seen hobbyists pass over a coin only a few inches deep, yet others use the same detector to detect coins at extreme depths. A most important factor, therefore, in successful detector operation is the expertise and ability of the operator.

There is no doubt in my mind that as long as it performs basic functions properly, even a poorly built detector will produce more in the hands of an experienced hobbyist than a high quality detector will in the hands of a person who does not understand the instrument or is not willing to learn how to use it.

This book will help *you* develop that experience!

Choosing a Detector

When selecting your detector, I urge you not to make the common mistake of thinking that you must choose the highest priced detector to get the best instrument...or, think that you must have the fanciest detector to get the most out of the hobby. Even if you can afford it, you may find yourself with an instrument that's so complicated that you'll spend more time adjusting controls and loading programs than treasure hunting!

Remember, you're looking for value. That means the "most detector for the dollar." Perhaps you've heard that metal detectors are expensive. Well, they do cost more than a lot of other hobby-type items. There's no question about that. But, remember, you must have a quality detector to appreciate the hobby, and we're talking about the *one-time* purchase of a scientific, electronic instrument. Plus, I personally guarantee that if you select metal detecting as a hobby, you won't ever be forced to buy bait, lures and balls or to pay greens fees, club dues or marina rentals like you would with some other, far more expensive, pastimes.

What Should It Cost?

How much should you spend for a detector? Remember that you generally get what you pay for. Because a quality metal detector is really a scientific instrument, you shouldn't expect it to be inexpensive. High quality detectors may seem expensive, but you'll never "lose your money" by purchasing one of them because you'll always have a high quality detector. On the other hand, if you buy a cheap, off-brand model, you'll essentially lose *all* of your money because your so-called detector will scarcely detect at all or do any of the other things you expect. You're left with nothing! In fact, starting out with a cheap detector is a sure-fire way to shorten your detecting career. You'll quit in a hurry because you won't find anything.

And, unfortunately, you'll probably tell people, "Oh, those metal detectors don't work. I have one that won't find anything!" Ever hear that? I sincerely believe that cheap detectors have done more harm to our hobby than just about anything else.

Let me urge you not to buy a model from any manufacturer that has a list price of much less than $200. And, remember, a quality metal detector will literally last a lifetime. The first one I ever made more than 30 years ago is now in our Garrett Museum, but I could take it into the park today and find coins.

Over the years my company has designed and built a wide range of detectors, and we've sold lots of each model. Our detector line includes different kinds of instruments with different features. Yet one of our continuing goals is to give every customer the most *value* for his or her money — no matter what Garrett detector he or she buys. We've even finally achieved my personal goal of a simple OneTouch computerized instrument that beats that $200 price.

Keep It Simple

A popular adjective for detectors today seems to be "simple." A competitor describes his instrument as "the ultimate in simplicity." Believe me, it isn't. Don't let yourself be taken in by advertising claims. Make certain that the detector you buy is ready to hunt when you turn it on. You don't want to have to set several controls or — perish the thought — load a program. I don't know about you, but having to *load* a program into my detector or to remember numerical or graphic codes to understand its signals ranks right up there with programming your VCR or computer in simplicity. "What you see is what you get" should be your goal in purchasing a OneTouch detector.

In addition to the OneTouch feature you should select a computerized detector — one with microprocessor controls. As we approach the 21st Century there's no reason you should be hunting with an instrument whose circuitry design is 25 years old. Another feature you should look for in a detector is a visual system of target identification and discrimination such as that found in a GTA instrument. It helps you understand what your detector is trying to tell you. Depth measurement and automatic pinpointing are similarly helpful. Headphone capability is a must since you'll probably be using them most of the time.

Select a detector built by a progressive company with a history of engineering excellence...a company that has steadily introduced quality improvements. Does the manufacturer test his own instruments? Does he get out into the field and use them under all kinds of situations? Does he travel to various locations to test varying soils to insure that his detectors work regardless of conditions? Are company engineers active in the field?

Choosing the detector that's just *right* for you can be a major factor in your enjoyment and appreciation of treasure hunting. If you're not happy with the instrument you're using, you're not going to get all the pleasure possible out of this great hobby.

And, what a tragedy that would be!

Still, the amount of success you ultimately obtain with any kind of metal detector (or with any other hobby) will be in direct proportion to the amount of time and study that you devote to it. So, let's make review the basic techniques of finding treasure with a modern metal detector.

The first instruction is: *Read your instruction manual!* Carefully study the operator's manual that accompanied your detector. At our Company we recommend that you examine this manual *before* you purchase a detector. If the manufacturer has "skimped" on providing instructions and advice, you might find yourself shorted in other areas of an instrument.

On the other hand, an extremely detailed manual that you find hard to understand could well indicate a detector that's going to prove just as complicated...one that will be difficult to learn how to use. New OneTouch computerized and microprocessor-controlled detectors are *simple to use*, not difficult.

If you purchased a used detector with no manual or the instrument was given to you without a manual, call the manufacturer immediately. I don't know about other companies, but Garrett will provide for only a nominal charge a manual for any detector we ever manufactured.

Getting Started

With a OneTouch detector you're ready at once to hunt for treasure. And, that's what you should do!

1. With the searchcoil held a foot or two above the ground touch one control, and you're ready to hunt. Lower your searchcoil to a height of about two inches above the ground and begin scanning.

2. Keep the searchcoil level as you scan and always scan slowly and methodically; scan the searchcoil from side to side and in a straight line in front of you with the searchcoil skimming the grass slightly above the ground. Do not scan the searchcoil in an arc unless the arc width is narrow (about two feet) or unless you are scanning extremely slowly. The straight-line scan method allows you to cover more ground width in each sweep and permits you to keep the searchcoil level, especially as you end each sweep. This method reduces skipping and helps you overlap more uniformly.

3. Overlap by advancing the searchcoil as much as 50% of the coil's diameter at the end of each sweep path. Occasionally scan an area from a different direction. Do not raise the searchcoil above scanning level at the end of each sweep. When the searchcoil begins to reach the extremes of each sweep, you will find yourself rotating your upper body to stretch out for an even wider sweep. This gives the double benefit of scanning a wider sweep and gaining additional exercise.

Over

Although, designed to hunt for gold, the Scorpion Gold Stinger with its famed Groundhog Circuit can also be used to detect coins and other treasures.

Facing

The GTI 2000 with Imaging by Garrett represents the latest innovation in metal detectors and will report the size and depth of every target.

4. As you scan the searchcoil over the ground, scan at a rate of about one foot per second. Don't get in a hurry, and don't try to cover an acre in 10 minutes. Always remember that what you are looking for is buried just below the sweep you are now making with your searchcoil. It's not across the field.

Finding a Target

When the sound increases and/or an indication is shown visually, a target is buried in the ground below the searchcoil. Acceptable objects will cause the detector's audio to get louder, and meter indicators will increase in amplitude.

Meters and LCDs can provide additional information concerning the possible "value" of targets. Garrett's LCD, called a Graphic Target Analyzer™, will show you every target — even the "bad" ones for which it makes no sound.

Over

Garrett's CX line of OneTouch detectors offers reliable, deepseeking all-purpose excellence for every treasure hunting task.

Facing

Metal detector evolution has resulted in Garrett's Treasure Ace, which offers computerized hunting with microprocessor controls at economy prices.

And, the GTA on this precision instrument actually indicates what you've discovered. That's right, there's a Target ID Guide located above the LCD. With a GTI detector you'll even be given an indication of the *size* of your target. These are just ways in which the GTA and GTI detectors have revolutionized treasure hunting.

As detectors continue to improve, even more target information will be presented visually.

Remember that a quality detector will never "lie" to you. It will simply report what is beneath its searchcoil. It's up to you to interpret this information. And, this will be a snap if you're using a detector with a system of visual identification. A glance at its LCD and Target ID Guide should give you a pretty good idea of what you've found...even its size if you're using a GTI.

Next, locate your target precisely by pinpointing. You accomplish this by scanning back and forth over the spot where you got a signal. Draw an imaginary "X" over this spot and dig there. Some detectors offer a mode called automatic pinpointing that aids in this effort. When automatic pinpointing is available, it enables the instrument to hover directly over a target with no motion necessary. Many hobbyists, especially old-timers, prefer to pinpoint their targets manually, even when their detector offers the automatic feature.

How do you dig a target? This is discussed in Chapter 3. Always remember, however, to make as small a hole as possible and to fill in your hole after you dig a target. Holes are not only unsightly, but they can be dangerous to people walking in the area. Perhaps it might be you! Before filling a hole, however, be sure to check it again with your detector to make certain you have recovered everything in and around it. It's embarrassing to have someone recover a target in the loose dirt of a hole you originally dug and filled. I know; it's happened to me!

Don't be concerned — at least at first — about how often you're discovering targets or just *what* you are finding. As you recover various targets, you will find yourself getting better and better with your detector. You will become more at ease in using it, and the quantity of found items will be growing at an accelerated rate.

If you haven't started using headphones, now is the time to do so. You'll learn how important they really are. You'll hear sounds you didn't hear before and find objects that you couldn't detect just by listening to a detector's speaker.

As you put miles behind your searchcoil, you will find yourself getting better and better with your detector. You will become more at ease in using it, and there will be fewer and fewer "problems" that bother you. The quantity of found items will be growing at an accelerated rate. All though your learning and training period and even on down through the years, you must develop persistence.

After you have some experience and get more comfortable with your detector, it's time to go back over the same areas you searched before you learned how to use your machine. You'll surprised at the quantity of coins and other objects you missed. In fact, each time you come back to these places you'll find more coins and other treasures, especially at greater depths, and experience fewer problems using the instrument.

And, speaking of problems, run-down batteries are by far the single most common source of detector "failure;" be sure to check your batteries before venturing out, and carry spare batteries whenever you are searching.

Increasing Your Enjoyment

That's all you need to know to find treasure with a OneTouch detector. But, you'll probably want to become familiar with all of your instrument's functions and learn more about all that it can do for you.

So, it's back to the instruction manual. You can skip the parts that concern assembly and searching, but read the rest of it again...this time, even more carefully. If you detector permits regulating the sound threshold, you'll want to learn how to do it. Follow instructions to set a minimum threshold level. If silent operation is desired, always make certain that such operation is just below an audible level. Many of us never use silent audio since it is possible to overlook "fringe" signals and miss targets.

The instruction manual and video should guide you through other adjustments which you can make through the bench-testing process. Lay your detector on a wooden bench or table. Do not use a table with metal legs and braces because the metal could interfere with your testing. Begin with the part of your instruction manual that describes detection of metal. Go through the procedure! If your detector is equipped with a sensitivity or detection depth control, test the detector at several levels. Test the instrument with various metal targets. Make the adjustments discussed in the manual or video that let you accomplish different objectives with the detector. Try as many of these variations with different targets on the bench as possible.

Then, you can experiment with the discrimination capabilities that will be provided by any quality detector. In the beginning, however, I suggest you use the basic mode that any modern computerized detector will offer when first turned on.

After you have used your detector for several hours, you can begin to test its discrimination capabilities for yourself. Read about this in your owner's manual. But, whatever you do, don't use too much discrimination...just enough to eliminate from detection the junk you may have been digging. Many modern computerized instruments offer visual discrimination that lets you *see* which types of targets you are asking your not to detect. Yet these same modern detectors will report visually on *every* target they encounter. They just won't announce audibly the ones you've chosen to discriminate.

Make Your Own Test Plot

After your interest in the hobby grows, you might want to construct your own test plot to help learn the capabilities of your detector. Success in scanning here will be a measure of how well you are progressing and how well you have learned your equipment. First of all, select the area for it and scan this area thoroughly with no discrimination so that you can remove *all* metal from the ground. Select targets such as various coins, a bottlecap, a nail and a pulltab. Select also a pint jar filled with scrap copper and/or aluminum metal and a large object such as a gallon can. Bury all these objects in rows about three feet apart and make a map showing where each item is buried. Be sure to note its depth.

Bury pennies at varying depths, beginning at one inch. Continue, with the deepest buried four to six inches deep. Bury one at about two inches but stand it on edge. Bury a penny at about two inches with a bottlecap about four inches off to one side. Bury the bottlecap, nail and pulltab separately about two inches deep. Bury the jar at ten inches to the top of its lid. Bury the gallon can with the lid 18 inches below the surface.

The purpose of the buried coins is to familiarize you with the sound of money. If at first you can't detect the deeper coins, don't worry. After a while, you'll be able to detect them quickly. Trust me. When you can detect everything in your test plot with ease, rebury some of the items deeper. The penny buried next to the bottle cap can give you experience finding a good target next to trash and will help you learn to distinguish individual objects. The jar and gallon can will help you learn to recognize "dull" sounds of large, deeply buried objects. Check the targets with and without headphones. You'll be amazed at the difference headphones make.

From time to time you may want to expand your test plot, rebury the targets deeper or experiment with new ones. Always leave an area totally void of targets. You will want to bury new ones from time to time to see how they sound or simply to test out your detection ideas. Remember to make an accurate map and keep it up to date when you change and/or add to your test plot.

Miscellaneous Tips

When searching areas adjacent to wire fences, metal buildings, metal parking meter posts, etc., reduce detection depth and scan the searchcoil parallel to the structure. This lets you get as close to it as possible.

Coins lying in the ground at an angle may be missed on one searchcoil pass but detected when the searchcoil approaches from a different angle.

If your detector has a volume control, keep it set at maximum. Don't confuse volume with audio (threshold) control. You may want to use a set of headphones with individual earphone volume adjustment and set each one to suit yourself.

Use your common sense. *Think* your way through perplexing situations. Remember, expertise is gained through research, patience, enthusiasm and the use of common sense.

Don't expect to find tons of treasure every time you go out! In fact, there may be times when you don't find anything. There are times when I don't. But the hobby's real joy and the reward of detecting is never knowing what you'll dig up next!

Success stories are written every day. A lot of treasure is being found and a lot of treasure is waiting to be found where you live. Detectors are not magic wands, but when used correctly they will locate buried and concealed treasure. Use a high quality detector and keep your faith in it. Have patience and continue using your instrument until you have it mastered. Success will be yours!

What a wonderful hobby this is! And, to make it even more wonderful we should always consider the important aspects of health, safety and the law, all of which we'll discuss in the following chapter.

Use A Detector;
Stay In Shape

Cabinetmaker Nyal Thomas got into the metal detecting hobby years ago by building a replica of a pirate's treasure chest and trading it to Charles Garrett for a metal detector. The swap was a good one; the chest is still in the Garrett museum, and Nyal found countless treasures (including a gold coin) with that first Garrett detector and others that he purchased after becoming a dedicated hobbyist.

As Nyal grew older, however, he engaged in other pursuits and allowed his interest in treasure hunting to wane. Yet, love for metal detecting was not forgotten, and he occasionally took his detector on nightly walks.

Well, he's now a full-time hobbyist again because he recognizes how important metal detection can be for older men and women. "After I had stooped to pick up a coin the other night, I could hardly get up. Why, if I'd had a phone, I'd have called 911. It was clear what good physical shape metal detecting had kept me in over the years."

Nyal immediately purchased a GTA and revived his interest in the hobby. His experience and computerized circuitry of the new detector resulted in immediate success. In fact he's already considering the purchase of an even better Garrett because of all that treasure hunting with a metal detector can do for him.

In Closing

*A*lthough the three subjects covered in this chapter should always be considered, I urge you never to let any of them become obstacles to your enjoyment of the hobby…or to use one as an excuse for not participating in it fully. Because you are aware of your health and safety, you conduct yourself accordingly.

Plus, I've always tried to abide by the maxim of that popular old cartoon character, Li'l Abner, who said, "I obeys every law, whether they're good or bad!"

Health

Major benefits to be derived from treasure hunting, concern the *health* of the hobbyist — mental as well as physical! Good exercise outdoors in the fresh air…exercise that is sustained but not overly strenuous…exercise under the absolute control of the individual, if you will, benefits men and women, girls and boys, of all ages. And, the zest and thrills that the hobby can bring are an absolute joy to the soul. There's no "time limit" to hunting with a metal detector, and a hobbyist is never forced to "keep up" with a younger, more athletic or experienced competitor. Alone or with others, anyone can hunt for treasure for hours a day or for just a short while. The effort may be intense or involve little exertion.

I have been hunting with metal detectors for most of my adult life and have never had any serious physical problems. I develop more aches and pains from using gym equipment. Over the years, I have developed four ways to lessen the dangers of strained or sore muscles:

– Select proper equipment, including accessories. This particularly concerns the detector's stem…if it is too long you will have a balance problem; too short, and you'll have to stoop over to search.

– Strengthen hand, arm, back and shoulder muscles with a regular, planned exercise program. Not much is really required here…in fact, just using a detector will probably develop the proper muscles. At the beginning, or after a period of inactivity, however, a hobbyist should protect against strained muscles and ligaments.

– Warm-up exercises before each day's activity are generally the answer. Just a few minutes of stretching and other activity to loosen muscles and joints will prepare them for a day's work.

– Finally, during metal detecting activities, use correct scanning techniques and follow accepted rules for stretching, bending and lifting. Whenever you feel yourself tightening up, take a short break. Most likely, however, stopping to dig targets will provide the rest you need.

Most important of all, use common sense and take care of yourself! There are no "time limits" to metal detecting. You have the rest of your life.

Safety

Hunting with a metal detector is generally as safe as any other outdoor hobby. Actually, from the standpoint of health and safety the worst things that will probably ever befall a hobbyist are sunburn or getting wet in a sudden storm. Even these can be minimized by sunscreen, proper clothing and following common sense rules of exposure.

You must decide for yourself how safe you will be in the areas where you plan to search. Accurate knowledge will not only help you dispel many unreasonable fears, but can materially reduce the chances of encountering problems. It is the *unknown* that we fear most. The best way to avoid trouble is to be ready for it at all times. Remember the Boy Scout motto: Be Prepared.

If you're in the wrong place at the right time, there's always the chance of your being bothered by someone. If you lack confidence in the security of an area, work in pairs. Some hobbyists exploring deserted areas carry a can of "mace" or similar deterrent...not stored away in a bag, but where it is readily accessible to them. I suggest you never tell anyone, even children, the amount of treasure you are finding. The quickest way to discourage people is to show them a few pulltabs and bottlecaps. They'll suddenly lose interest and even the children won't be so anxious to help you dig. Never tell inquisitive people how much your detector is worth. Just say, "Oh, they don't cost very much; besides, this detector was a gift." In fact, it probably was a gift, either from yourself or from your spouse.

Always be alert to the possibility of digging up explosives. Over the past half century some areas have been used from time to time as bombing and artillery ranges. Now, these areas are certainly few and far between. Nevertheless, if you dig up a strange-looking device that you suspect might be explosive, notify the authorities immediately. Let them take care of it. Then exercise caution when digging in that area, or just stay away entirely. The same advice applies to underground cables or pipelines.

Watch where you're walking! Of course, you won't fall in the holes you dig, certainly, but joggers and others might, if you fail to cover them. So...*fill your holes!*

Many natural sites represent a fragile environment that can be easily damaged or destroyed. Please leave only footprints — not pulltabs, wrappers, cans or other souvenirs of our "disposable" civilization. Remember, a fellow treasure hunter may want to work the area someday. *You may even want to come back yourself!*

When campfires are covered and not doused with water, coals remain very hot even till the next day and can cause severe burns. Watch out for coals, even when they appear cold.

Toxic waste presents an increasingly serious problem. Be alert to any area (or any piece of flotsam or jetsam on a beach) that looks or smells bad...in any way. *Keep away from anything* that you suspect of being contaminated.

Probably the greatest danger facing anyone in strange places is panic. Such a feeling can easily occur in any deserted area, particularly if you are alone. Sudden overwhelming fear, accompanied by loss of reasoning, contributes to a great many accidents. Condition yourself to resist panic. Try to think calmly about each problem, even before you face it. Acting quickly without thinking can usually gains you nothing. Fear can be overcome. Let your reasoning take control to allow you to *think your way* out of difficult situations.

Don't Get Stung!

The above warning doesn't concern buying a low quality detector...that's discussed elsewhere! It's inevitable that people who venture outdoors are going to encounter gnats, mosquitoes, bees, ants, wasps, spiders, ticks, hornets, scorpions and just plain bugs. You scarcely need to be told to try to avoid them. Even when you can't, however, the worst that usually results is a brief moment or so of slight discomfort. But, the results have been known to be much worse. I'm certainly not trying to scare you with talk about insects, spiders and such. But, it's a good idea to keep them in mind any time you're out of doors.

Above all, use your good common sense, and you'll be fine! Never let needless worry interfere with the joy and thrill of this great hobby. Accurate knowledge will not only help you dispel any unreasonable fears, but materially reduce the chances of encountering problems.

The Law

As a hobbyist looking for coins in the park or on property belonging to an individual who gave you permission to hunt, you aren't going to run afoul of any laws that govern hunting for historic artifacts or disturbing potential archaeological sites. I would like to raise just a few legal points, however, that you should consider before going out into the field to scan with a metal detector and to remind you that there are laws applicable to various treasure hunting situations. Each state has its own statutes concerning where you can hunt and whether you may keep treasure when it is found. You should learn these laws.

Areas such as national and state parks and monuments must be considered as absolutely "off limits" to metal detecting unless you have been informed otherwise by the *proper authorities*. Don't rely on gossip. In fact, if you're really concerned about hunting in an area, it might be a good idea to get your permission in writing.

All states have laws against trespassing. If a sign says, "Keep Out," do just that. It is always best to seek permission wherever you hunt. Anyway, how can you listen to your metal detector if you have to keep an ear cocked for a returning property owner...or, a siren?

Ownership of Property

Finder's Keepers — There may be some truth in this old statement, especially about unmarked items such as coins. But, there are certainly exceptions, particularly when you start considering other objects whose ownership can be more easily identified. No matter what kind of treasure you are looking for, I urge you to have a general knowledge of the laws of ownership. You can never tell what you'll find or where you'll find it! Finder's Keepers may not be appropriate for an object you discover on private or posted property if the landowner decides to dispute your claim. On the other hand, Finder's Keepers generally applies to any owner-not-identified item you find when you are not trespassing, when you are hunting legally on any public land and when the rightful owner cannot be identified.

Treasure trove — In the United States this is broadly defined as any gold or silver in coin, plate or bullion and

This thrill of discovery is the common goal of countless thousands of men and women of all ages who seek treasure with a metal detector.

to indicate that the original owner is dead or unknown. All found property can generally be separated into five legal categories:

Abandoned property, as a general rule, is a tangible asset discarded or abandoned willfully and intentionally by its original owners. An example would be a household item discarded into a trash receptacle. If anyone decides to take the item, they can do so legally.

Concealed property is tangible property hidden by its owners to prevent observation, inventory, acquisition or possession by other parties. In most cases, when such property is found, the courts order its return to the original owner.

Lost property is defined as that which the owner has inadvertently and unintentionally lost, yet to which he legally retains title. Still, there is a presumption of abandonment until the owner appears and claims such property, providing that the finder has taken steps to notify the owner of its discovery.

Misplaced property has been intentionally hidden or laid away by its owner who planned to retrieve it at a later date

Charles Garrett, searching here in the Robbers Roost area of Utah where Butch and Sundance hid out, has hunted literally all over the world.

but forgot about the property or where it was hidden. When found, such property is generally treated the same as concealed property with attempts required to find its owner. When this is not possible, ownership usually reverts to the occupant or owner of the premises on which the misplaced property was originally found.

Things embedded in the soil generally constitute property other than treasure trove, such as antique bottles or artifacts that might be of historical value. The finder acquires no rights to the object, and possession of such objects belongs to the landowner unless declared otherwise by a court of law.

With the proper attitude and a true explanation of your purpose, you will be surprised at the cooperation you will receive from most landowners. The majority of them will be curious enough about your metal detector and what you hope to find, to agree to let you search. Offer to split, giving them 25% (or less) of all you find and they will usually be even more willing. If large amounts of treasure are believed to be hidden or buried, a properly drawn legal agreement is a *must!* Such an agreement between both you and all landowners (husband and wife, etc.) will eliminate any later disagreements which might otherwise arise.

You will probably be pleasantly surprised at how much public property is open for you to search with a metal detector to your heart's content. Yet, believe me when I say that these public areas will not remain open if we hobbyists do not behave properly as we search for treasure. Do not damage the grass or shrubbery or leave behind trash or holes. All hobbyists must become aware of their responsibility to protect the property of others and to keep public property fit for all.

Rules of Conduct

Of course, the first rule of conduct for any treasure hunter is to *fill all holes*. You'll learn that most every governmental subdivision — be it city, township, county, state or whatever — enforces some sort of law that prohibits destruction of public or private property. When you dig a hole or cut through the grass on private or public property, you're in effect violating a law. Of course, laws are generally not enforced this rigidly, especially if the hobbyist is careful in his digging and retrieving.

Of course, there have always been laws to protect private as well as public property, but only in recent years have these been rigidly enforced to limit the activity of metal detector hobbyists. Why has this happened? Public lands, parks, recreational areas and such are continuously maintained and kept in good condition so that those using such facilities can enjoy them to the fullest. When there is willful destruction, laws protecting the property are more rigidly enforced and new laws are sought. There are numerous methods you can use to retrieve coins and other objects without destroying landscaping and making unsightly messes.

In addition, property should always be restored to the condition in which you found it. I have heard of so-called treasure hunters who completely devastate an area, leaving large gaping holes, tearing down structures and uprooting shrubbery and sidewalks. Damage of this kind is one of the reasons we're seeing so many efforts at legislation that would literally *shut down* metal detectors on public property.

An experienced hobbyist always seeks to leave an area where he or she has hunted in such a condition that nobody will know that it has ever been searched. In fact, I always urge hobbyists to leave any area they explored in *better* condition than they found it! All treasure hunters must become aware of their responsibilities to protect property of others and to keep public property fit for all. Persons who search for valuables by destroying property, leaving holes unfilled or tearing down buildings should not be known as treasure hunters but should be called what they are — looters and scavengers!

Code of Ethics

Filling holes to protect the landscaping is but one requirement of a dedicated metal detector hobbyist. Thousands of individuals and organizations have adopted a formal Metal Detector Operators Code of Ethics:

"— I will respect private and public property, all historical and archaeological sites and will do no metal detecting on these lands without proper permission.

"— I will keep informed on and obey all laws, regulations and rules governing federal, state and local public lands.

"— I will aid law enforcement officials whenever possible.

"— I will cause no willful damage to property of any kind, including fences, signs and buildings, and will always fill holes I dig.

"— I will not destroy property, buildings or the remains of ghost towns and other deserted structures.

"— I will not leave litter or uncovered items lying around. I will carry all trash and dug targets with me when I leave each search area.

"— I will observe the Golden Rule, using good outdoor manners and conducting myself at all times in a manner that will add to the stature and public image of all people engaged in the field of metal detection."

Taxes

All treasure that you find must be declared as income during the year in which you receive a monetary gain from that treasure. If you find $1,000 in coins, which you spend at once because they have no numismatic value, then you must declare the face value of those coins in the current year's income tax report. If, however, you discover a valuable coin — or, say, an antique pistol — you do not make a declaration until you sell the item(s) and then only for the amount you received. If you decide to donate some of your finds to historical societies or museums, you may be able to deduct the fair market price of the items as charitable contributions. Simply stated, the tax laws require you to declare all income from treasure hunting.

You may be allowed to deduct some or all of your expenses but you must have good records. You are advised to check with an attorney or tax accountant, especially if you plan to become a full-time treasure hunter. An accountant will advise you as to what type records you should keep.

Last, But Not Least

Our hobby — the sport of searching for treasure with a metal detecting — has been kept clean and dignified by people who care about it, while they express a similar concern for themselves and their fellow man. Most of us who use metal detectors will go out of our way to protect this most rewarding and enjoyable hobby that we love so much...as well as share our enjoyment with others. Why, the simple act of sharing is the primary reason that I write books such as this. I *enjoy* helping others to find treasure because, as famed underwater salvor Bob Marx likes to say, "There's plenty for all!"

Yet, keeping the hobby clean takes the effort and dedication of everyone...not just a few. So, as you go about enjoying your leisure — or perhaps full-time (lucky you) — activity, be professional! Be worthy of this great hobby!

I wish you every success and great happiness. And, I sincerely hope that someday, when we're both out searching...

I'll see you in the field!

Terms & Techniques

*S*ure, we can always learn more, but this book has already told you all you really need to know for a happy and successful pursuit of the hobby of treasure hunting with a detector. I could have entitled this concluding chapter "Glossary," but it's really more than that. You read in Chapter 12 about metal detector terminology, and you'll find many of those terms in this chapter. But it also provides additional information on the Terms and Techniques as they relate to treasure hunting with a metal detector.

A newcomer to our hobby might occasionally hear a word or phrase with which he or she isn't familiar and wonder what it means. Of course, you want to be knowledgeable about your hobby, and this chapter will help you. But, remember...some of these terms have significance only from a *historical* standpoint. They do *not* apply to a modern OneTouch computerized detector. You'll probably want you to know what they mean, but I want you also to understand that they have no pertinence to today's detectors or to our modern hobby — no matter what any so-called "professional" treasure hunter might try to tell you!

Italicized words within a definition refer to another defined term.

Air Test — A method, once somewhat valid, to determine the *Sensitivity* of a metal detector; i.e., how deeply it can detect. So called, because the test is performed with nothing but air between the detector's searchcoil and the object being detected. Depending primarily on soil/mineral and atmospheric conditions but also on the detector itself, depth performance in the field can (and, usually does) vary widely from that of an air test. Computerized detectors can usually detect deeper in the field than in an air test.

All Metal Mode — Another name for the *Deepseeking* mode of detector operation in which all metal targets are detected.

Ampere — A unit of electrical current which measures rate of flow of electrons in a conductor.

Amplifier — An electrical circuit that draws its power from a source other than the input signal and which produces an output voltage/current that is an enlarged reproduction of the essential features of the input signal.

Antenna — The component of a transmitter or receiver that actually radiates or receives the electromagnetic energy. (See *Searchcoil.*)

Audio Adjust — The control used to adjust the sound produced by the speaker or *Headphones* of a metal detector. Not necessary on most modern detectors. (See *Tuning* and *Volume.*)

Automatic (Audio) Tuning — A circuit incorporated in most modern detectors that keeps the audio *Threshold* level at a predetermined setting.

Automatic Ground Balancing — Circuitry featured on most modern discriminating *Metal Detectors*, requiring no manual adjustments to cancel out detrimental effects of iron earth and salt mineralization. Some detectors in their Deepseeking all metal mode also offer this feature. (See *Ground Track.*)

Automatic Pinpointing — An electronic aid to precise target location that causes a "sharpening" of detector signals when objects are detected. .

Bench Test — Static assessment of capabilities of a *Metal Detector*, usually lying on a bench, table or other surface.

BFO Detector — An obsolete type of metal detector utilizing Beat Frequency Oscillator (BFO) circuitry popular in the 1960s and 1970s. Results possible with such detectors are totally unsuitable when compared with the capabilities of modern instruments.

Black Sand — See *Magnetic Black Sand.*

Body Mount — (See *Hip Mount.*)

Calibration — A term that generally refers to factory adjustment of a detector to specific operating performance. For instance, accurate ore-sampling requires that a detector be set at a factory-calibrated point at which the distinction between metal and mineral is clearly recognizable.

Canceling — Obsolete term sometimes used to describe *Ground Balancing* or *Discrimination.*

Circuitry — An electrical or electronic network providing one or more closed electrical paths. More specifically, it is a grouping of components and wiring in devices designed to perform some particular function or group of functions. Examples of circuits within a metal detector are transmitter circuit, receiver circuit, antenna circuit and audio amplifier circuit. (See *Microprocessor.*)

Circuit Board — The thin sheet of material upon which

electronic components are mounted If the circuit board is completely self-contained, it is usually referred to as a "module." Circuit boards may be hand-wired or have the interconnectors printed electrochemically upon them. Such modules are then designated PCB (printed circuit board). (See *Surface-mount PCB*.)

Classification, Audio — A method (or methods) in which a *Metal Detector* classifies detected targets into conductivity classes or categories with audible signals. (See *Coin Alert*.)

Classification, Visual — A visual (metered or light) method(s) for classifying detected targets into conductivity classes or categories. (See *Meter* and *Graphic Target Analyzer*.)

Coil—See *Searchcoil*.

Coin Alert™ — An audible method of producing a special tone only when coins (or high conductivity silver and gold items) are detected. All other detected targets produce normal accept/reject signals. A Garrett trademark.

Component — In a metal detector this term generally refers to an essential part of a circuit; i.e., resistor, capacitor, coil, tube, transistor, etc. The term can also refer to complete functional units of a system; i.e., transmitter, receiver, searchcoil, etc.

Computerized Detector — A *Metal Detector* that utilizes an integrated circuit containing the necessary elements of a small digital computer. This *Microprocessor* chip gives the detector "memory" that enables it to perform automatically numerous functions that are the responsibility of the operator of a non-computerized detector, if the functions are indeed ever performed at all.

Conductance — The ability of an element, component

or device to permit the passage of an electrical current; i.e, *Eddy Currents*. It is the reciprocal function of resistance.

Conductor — A wire, bar or metal mass (coin, gold nugget, ship's hull, etc.) capable of conducting electrical current.

Control Housing — The container in which all or most of the electronic assembly and batteries of a metal detector are placed. Modern detectors use high strength engineering polymers for ruggedized housings that are compact and stylish.

Deepseeking Mode — Mode of detector operation in which all metal targets are detected and greatest depth for all target sizes can be achieved. Often used for the first scan of a site to detect objects that the *Discriminate Mode* might be set to reject. The *Discriminate Mode* is then used to identify targets, permitting the operator to decide whether to dig or not. Deepseeking Mode is preferred by gold and cache hunters.

Depth Detection — A term usually used to describe the ability of an instrument to detect metal objects to specific depths. (See *Sensitivity*.)

Depth Penetration — Applied to electronic metal detectors, the term is used to define the specific distance into a particular medium (soil, water, air, etc.) that the electromagnetic field of a metal detector is capable of satisfactorily penetrating. This affects the *Depth Detection* abilities of an instrument.

Depth Scale — The markings (or LCD indicators) on a *Meter* or *Graphic Target Analyzer* that report the depth of coin-sized targets.

Detection Pattern — See *Searchcoil Detection Pattern*.

Discriminate Mode — The mode of operation of a *Metal Detector* in which the operator determines which metallic targets are to be detected. (See *Discrimination*.) It is the only mode of operation offered by a Discriminating Detector, and is a secondary mode offered by quality detectors with a *Deepseeking Mode.*

Discriminating Detector — One of today's most popular type of instruments, especially for hunting coins and searching beaches.

Discrimination — The ability of circuits within a detector to eliminate from detection metallic objects with specific conductivities. Using a *Discriminating Detector* an operator chooses which types of targets are to be detected through use of simple discrimination control(s). This function is sometimes described as *Elimination.* (See *Notch Discrimination*.)

Eddy Currents — Also called Foucalt currents, they are induced in a conductive mass by the variations of electromagnetic energy radiated from the detector and tend to flow in the surface layers of the target mass. Flow is directly proportional to frequency, the density of the electromagnetic field and the conductivity of the metal. They are a primary electrical phenomenon that produces detection signals in all metal detectors.

Electromagnetic Field — An invisible field that surrounds the transmitter winding; generated by the alternating radio frequency current that circulates in the transmitter antenna windings.

Electromagnetic Induction — See *Induced Current.*

Electronic Circuit — A circuit wherein current flows through wires, resistors, inductors, capacitors, transistors and other components.

Electronic Prospecting — A term sometimes used to describe the use of a metal detector to search for gold, silver or other precious metals in any form. Any quality metal detector can be used to hunt for nuggets, but the most common electronic prospecting is the search for gold using a detector with a *Deepseeking All Metal Mode.*

Elimination — Terminology that is sometimes used to explain characteristics more accurately described as *Ground Balancing* or *Discrimination.*

Elliptical Searchcoil — A specially designed oval-shaped searchcoil with length approximately twice its width.

False Detection — Responses to objects or anomalies other than sought metallic targets.

Fast Track™ — A type of detector circuitry, requiring computerized microprocessor controls, that automatically *Ground Balances* a detector's *Deepseeking Mode* circuitry. A Garrett trademark. (See *Ground Balance* and *Ground Track.*)

Ferrous — Pertains to iron and iron compounds, such as nails, bottlecaps, cannons or ships hulls.

Gain — An increase in voltage, current or power with respect to a previous quantity or a standard reference. Gain occurs in vacuum tubes, transistors, transformers, etc. as gain per component, gain per stage and gain per assembly. Such gain can be measured in terms of voltage, current, power or decibels.

Graphic Target Analyzer™ — That device on a metal detector that reports continuously and visually on an *LCD* such information as depth and type of target, audio and tone levels, sensitivity, battery condition, etc. A Garrett trademark.

Graphic Target Imaging™ — That capability of a Garrett detector to indicate on its TreasureVision™ screen the approximate depth and size of a detected target.

Gravity Trap™ **Gold Pan** — The patented (U.S. Patent #4,162,969) gold pan featuring 90° Riffles made and sold by Garrett.

Ground Balancing — Elimination of the detection effect of iron minerals or wetted salt. Once a required skill in metal detecting, this action is now performed automatically by the circuitry of *Discriminating Detectors.* It should, therefore, be on no concern to most hobbyists who use modern computerized instruments. (See *Automatic* and *Manual Ground Balancing, Fast Track* and *Ground Track.*)

Ground Track™ — A type of detector circuitry, requiring computerized microprocessor controls, that automatically *Ground Balances* a detector's *Deepseeking Mode* circuitry and continually maintains proper ground balance while the detector is being *Scanned.* A Garrett trademark. (See *Fast Track* and *Ground Balance.*)

Hand-held Detector — A metal detector that can be held in the palm of an operator.

Headphones — Metal detector accessory that converts electrical energy waves into audible waves of identical form. Used by treasure hunters in place of detector loudspeakers, especially in noisy or windy locations.

Hip Mount — A detector configuration once popular because of the heavy control housings on older detectors. While this configuration is built into the Garrett's GTA detectors (belt-mounted battery pack) and is an option on the Scorpion Gold Stinger, most hobbyists no longer consider it important because of the light weight of modern quality instruments.

Imaging —The term used to describe a detector's ability to report on its LCD the relative size of any detected target. *Imaging by Garrett* indicates five distinct sizes, from "smaller than a coin" to "larger than a 12-oz. can."

Induced Current — The current that flows in a conductor or conductive mass when a varying electromagnetic field is present.

LCD — The letters LCD literally stand for liquid crystal display. They describe the constantly operating visual display of operational and target data on the *Graphic Target Analyzer* and *Treasure Eye* of Garrett detectors.

Magnetic Black Sand — Magnetite, a magnetic oxide of iron and, in a lesser degree, hematite; may also contain titanium and other rare-earth minerals but serves mainly as an indicator of the possible presence of *Placer* gold.

Magnetometer — Not a metal detector, even though the term is sometimes used improperly when metal detectors are discussed. Rather, this is an instrument for measuring magnetic intensity, especially the earth's magnetic field. Professional treasure hunters searching for large hidden masses of ferrous metal such as a ship or large cannon often use a magnetometer to locate the increased magnetic field density caused by the metallic mass.

Manual Ground Balance — Detector circuit in the *Deepseeking Mode* that permits precise personal selection of the normally automatic adjustments that eliminate the detrimental effects of iron and salt mineralization.

Matrix — The entire area below a searchcoil that is "illuminated" by the *Electromagnetic Field* transmitted from the antenna in the *Searchcoil.* A matrix may wholly, partially or intermittently contain conductive and/or non-conductive targets which may be of either ferrous or non-ferrous materials. The matrix may contain moisture, sulfides, metallic ores, etc. The *Detection Pattern* is only a portion of the matrix.

Metal Detector — An electronic instrument or device, usually battery-powered, capable sensing the presence of conductive objects lying underground, submerged within water, hidden on an individual or otherwise out of sight; then, providing its operator with an audible and/or visual indication of that presence.

Metal/Mineral — Refers primarily to that point in the *Deepseeking Mode* of a properly calibrated detector at which time any signal given by the detector will indicate that a target is metal, not mineral.

Meter — That device on a metal detector that reports information visually concerning depth and type of target, battery condition, ground conditions, etc.

Microprocessor — An integrated circuit that contains the necessary elements of a small digital computer. These circuits are now being used in the most advanced detectors. The "memory" of a microprocessor is preprogrammed to permit a detector to automatically perform numerous functions that even the most knowledgeable operator would normally have to carry out manually. In addition, microprocessor-controlled *Circuitry* performs these functions instantly and simultaneously as well as automatically.

Mode — The manner in which a detector operates which includes essentially the discrimination settings and other operating features required for seeking various types of targets such as coins, relics, caches, etc.

Narrow Scan — A scan width less than full searchcoil diameter. In earlier days, some searchcoils could scan an effective area equivalent only to about 30-40% of the diameter of the searchcoil itself.

Non-Ferrous — Pertains to non-iron metals and compounds, such as brass, silver, gold, lead, aluminum, etc.

Notch Discrimination — A special type of circuitry offered by some computerized microprocessor detectors that enables a detector to eliminate from detection a number of specific undesirable metallic objects that can widely differ in their conductivity. This enables a detector operator to be highly selective in choosing just which targets are to be sought and which are to be avoided. (See *Discrimination*.)

Null — A tuning or audio adjustment condition that results in "quiet" or zero audio operation.

Oscillator — The variation of an observable or otherwise detectable quantity of motion about a mean value.

PCB — Printed circuit board. (See *Circuit Board* and *Surface-mount PC Board*.)

Penetration — The ability of a detector to penetrate earth material, air, wood, rock, water, etc. to locate metal targets. Penetration is a function of detector design and type of detector, as well as the material being penetrated.

Performance — The ability of a detector to carry out the functions of which the manufacturer has claimed this instrument is capable.

Permeability — The measure of how a material performs as a path for magnetic lines of force as measured against the permeability standard, air. Air is rated as 1 on the permeability scale; diamagnetic materials, less than 1; paramagnetic materials, slightly more than 1; and ferromagnetic materials, much more than 1.

Pinpointing — The ability of a detector operator to

determine exactly where a detected target is located. (See *Automatic Pinpointing.*)

Placer — Pronounced like "plaster" without the "t," the term describes an accumulation of gold, black magnetic sand and other elements of specific gravity higher than sand, rock, etc. found in the same area.

Pulse Induction — Metal detection circuitry that employs the phenomenon (characteristic) of electromagnetic decay to sense the presence of a conductive material (metal). The system operates by delivering short bursts of energy to the antenna followed by a passive period when the antenna can sense the decaying electromagnetic field induced in a target. Garrett's XL500 Sea Hunter underwater detector utilizes the pulse induction method of detection.

Push-button — A type of mechanical control used in some, generally older, types of metal detectors. Distinctly different from the more modern electronic *Touchpads.*

Receiver — That portion of the circuitry of a metal detector that receives information created by the presence of targets, acts upon that information and processes it according to intentions indicated by instrument design or actions of its operator; then activates the readout system in proportion to the nature of the received data.

Receiver Gain — The amplification of input signals to whatever extent required.

R. F. — Stands for radio frequency; sometimes used to describe the transmitter frequency of the electromagnetic field being "broadcast" by a detector. Frequencies are commonly described as very low (VLF), low (LF), medium (MF), high (HF), very high (VHF), ultra high (UHF) and super high (SHF).

Scanning — The actual movement of a searchcoil over the ground or other area being searched.

Searchcoil — The component of a metal detector that houses the transmitter and receiver antennas and any associated circuitry. Usually attached to the control housing by way of an adjustable connecting stem, the searchcoil is scanned over the ground or other area being searched.

Searchcoil Detection Pattern — That portion of the *Electromagnetic Field* in which metal detection takes place. It is located out from and along the axis of the searchcoil generally, starting at full searchcoil width and tapering to a point at some distance from the searchcoil. Its actual width and depth depend upon the type and strength of signals being transmitted by the detector itself as well as the size and nature of any given target and the material(s) that compose the *Matrix*.

Sensitivity — The ability of a detector to sense conductivity changes within the detection pattern. Sensitivity of a detector increases inversely with the size of a metallic target that is detectable. One of the most important operational characteristics of an electronic metal detector, sensitivity determines the actual size of targets that an instrument will detect and the depth to which they can be detected. Although sensitivity is ultimately determined by design characteristics of a detector, a control on most instruments permits some regulation by the operator.

Signal — Generally describes the electromagnetic data received by the detector from a target and the audio and/or visual response generated by it.

Silent Audio — The tuning of the audio level of a detector in the "silent" (just below *Threshold*) zone. When operating in this fashion, the operator hears no sound at all until a target is detected. It is not a recommended method of operation, especially for a newcomer to the hobby.

Submersible — A designation of environmental protection indicating that complete submersion of a detector's *Control Housing* and/or *Searchcoil* will not affect its operation. All Garrett searchcoils can be submerged to the connector. The entire Sea Hunter detector can be submerged to 200 feet.

Super Sniper™ — A Garrett trademark used to describe its 4 1/2-inch searchcoil and the method for using it to enhance individual target detection in specialized situations. This type of treasure hunting is especially effective for hunting in areas with large amounts of metal "junk" or near metallic objects such as fences, posts, buildings, etc.

Surface Area — That dimension of a target lying parallel to the plane of the detector searchcoil; in other words, that part of the target that is "looking at" the underside of the searchcoil. This is the area through which the electromagnetic field lines pass and on which eddy currents are generated.

Surface-mount PC Board — Product of the modern method of manufacturing printed circuit boards (PCBs) in which automated techniques are utilized to mount miniaturized components on the board's surface for increased design efficiency, production economy and more reliable performance. (See *Circuit Board*.)

Sweeping — See *Scanning*.

Threshold — Adjustable level of audio sound at which a metal detector is operated when searching for treasure.

Touchpad — Type of control popular on modern detectors because of its effectiveness and dependability.

TR Detector — The type of metal detector that first utilized the Transmitter-Receiver circuit, which is essentially the circuitry of all modern instruments, except those with *Pulse Induction* circuitry. Development of TR detectors is important in the history of metal detector development, but results possible with the older detectors are totally unacceptable when compared with the capabilities of modern instruments.

Treasure Eye — That device on a metal detector that reports target location visually on an *LCD* and serves as an aid to *pinpointing* on certain of Garrett's Treasure Ace models.

Tuning — That adjustment an operator makes to bring the detector's audio level to a previously designated *Threshold* level. This adjustment is no longer of particular concern to hobbyists since it is accomplished automatically by the circuitry of modern detectors.

Universal Capabilities — Describes a metal detector that can effectively accomplish most THing tasks...coin hunting, beach and surf hunting, ghost towning, electronic prospecting, cache hunting, etc. (See *Versatility*.)

Versatility — A measure of the applications in which a detector can be used effectively. In other words, for how many different kinds of hunting can a particular detector be used? (See above.)

Visual Indicator — Generally means an *LCD* on a modern detector, although some older instruments utilize a *Meter*.

VLF Detector — The initials stand for Very Low Frequency, a segment of the R.F. spectrum that includes frequencies from 3 kHz to 30 kHz. At one time this designation was used to denote modern detectors, but almost all instruments (except Pulse Induction-types) now operate in this R.F. range. Some specialized detectors operate at frequencies up to about 100 mHz, but certain performance capabilities may be reduced.

Volume Control — A manual or automatic control used to limit voltage and/or current in an audio amplifier and, thereby, control volume of sound or "loudness" when a target is encountered. Do not confuse with the audio adjustments involved with establishing a *Threshold* setting.

Wetted Salt — The most prevalent mineral encountered in beach and ocean hunting; is essentially ignored by all *Pulse Induction* and those *Discriminating Detectors* with discriminating circuitry that cancels the conductive salt effect.

Wide Scan — Generally implies that the scanning width (detection pattern) of a detector, as the searchcoil passes over the ground, is equal to the full width (or wider) of the diameter of the searchcoil being used. This term is particularly appropriate to *Elliptical Searchcoils,* which can be designed to scan effectively over an area full length toe to heel. All detection patterns, however, taper to a point at some depth. (See *Searchcoil Detection Pattern.*)

Zero Discrimination — That adjustment in the circuitry of a detector when it provides no discrimination in a particular mode.

Form for Ordering...

Ram Books

Please send the following books:

- ☐ Treasure Hunting for Fun and Profit ....$ 9.95
- ☐ Ghost Town Treasures ....$ 9.95
- ☐ Real Gold in Those Golden Years ....$ 9.95
- ☐ Let's Talk Treasure Hunting ....$ 9.95
- ☐ Buried Treasures You Can Find...............$14.95
- ☐ The New Successful Coin Hunting ....$ 9.95
- ☐ Modern Metal Detectors $14.95
- ☐ Gold of the Americas $12.95
- ☐ Find Gold with a Metal Detector $12.95
- ☐ Gold Panning Is Easy ....$ 9.95
- ☐ New World Shipwrecks: 1492-1825 ....$16.95
- ☐ Treasure Recovery from Sand & Sea $14.95
- ☐ Sunken Treasure: How to Find It $14.95
- ☐ An Introduction to Metal Detectors ....$ 1.00
  (No shipping/handling charge for this book)
- ☐ Find an Ounce of Gold a Day ....$ 3.00
  (Included free with Garrett Gold Panning Kit)

Ram Publishing Company
P.O. Drawer 38649
Dallas, TX 75238
FAX: 972-494-1881
(Credit Card Orders Only)

Handling charges:
Please add $1
for each book
(maximum of $3)

Total for items $_____

8.25% Tax (Texas residents) $_____

Handling Charge $_____

                            TOTAL $_____

☐ Enclosed is check or money order

   I prefer to order through
☐ MasterCard
☐ Visa
By telephone:
1-800-527-4011   _____

                        Credit Card Number

_____

Expiration Date   **Phone Number (8 a.m. to 4 p.m.)**

_____

Signature (Credit Card orders must be signed.)

_____

NAME

_____

ADDRESS (For Shipping)

_____

CITY, STATE, ZIP